RISK MANAGEMENT

AND CONSTRUCTION

RISK MANAGEMENT
AND CONSTRUCTION

ROGER FLANAGAN

Professor of Construction Management
University of Reading

AND

GEORGE NORMAN

Professor of Economics
University of Edinburgh

OXFORD

BLACKWELL SCIENTIFIC PUBLICATIONS

LONDON EDINBURGH BOSTON

MELBOURNE PARIS BERLIN VIENNA

© Royal Institution of Chartered Surveyors
1993

Blackwell Scientific Publications
Editorial Offices:
Osney Mead, Oxford OX2 0EL
25 John Street, London WC1N 2BL
23 Ainslie Place, Edinburgh EH3 6AJ
238 Main Street, Cambridge,
   Massachusetts 02142, USA
54 University Street, Carlton
   Victoria 3053, Australia

Other Editorial Offices:
Librairie Arnette SA
2, rue Casimir-Delavigne
France

Blackwell Wissenschafts-Verlag GmbH
Meinekestrasse 4
D-1000 Berlin 15
Germany

Blackwell MZV
Feldgasse 13
A-1238 Wien
Austria

First published 1993

Printed and bound in Great Britain by
the University Press, Cambridge

DISTRIBUTORS

Marston Book Services Ltd
PO Box 87
Oxford OX2 0DT
(*Orders:* Tel: 0865 791155
        Fax: 0865 791927
        Telex: 837515)

USA
Blackwell Scientific Publications, Inc.
238 Main Street
Cambridge, MA 02142
(*Orders:* Tel: 800 759-6102
              617 876-7000)

Canada
Oxford University Press
70 Wynford Drive
Don Mills
Ontario M3C 1J9
(*Orders:* Tel: 416 441-2941)

Australia
Blackwell Scientific Publications Pty Ltd
54 University Street
Carlton, Victoria 3053
(*Orders:* Tel: 03 347-5552)

British Library
Cataloguing in Publication Data

A catalogue record for this book is
available from the British Library

ISBN 0-632-02816-5

Library of Congress
Cataloging in Publication Data

Flanagan, Roger.
    Risk management and construction/
Roger Flanagan and George Norman.
        p.   cm.
    Includes bibliographical references and
index.
    ISBN 0-632-02816-5
    1. Risk management. 2. Construction
industry.   I. Norman, George.
II. Title.
HD61.F53   1993
624'.068'4—dc20   93-20446
                                       CIP

CONTENTS

LIST OF FIGURES

LIST OF TABLES

FOREWORD

INTRODUCTION

We would like to acknowledge the support of The Education Trust of The Royal Institution of Chartered Surveyors who generously supported work in the area of risk management for construction. Our thanks go to the Steering Group who provided advice, enthusiasm and help in structuring the work.

There are many people to thank for their time and assistance when preparing a book. Everybody we approached was always objective, helpful, and enthusiastic. Space prohibits us from listing everyone, but we convey our thanks to the large number of people who gave us their time and helped to formulate ideas. We are grateful to Euro Log Ltd of Teddington for their enduring assistance and for allowing us to use the Case Study in Chapter 10. Our thanks also to John and Carol Jewell for their help in the production of this book.

All the shortcomings, omissions and errors are totally ours.

Risk management will continue to develop and every publication takes the subject a stage nearer a better understanding of the construction process. There is still a long way to go - there will always be risk in construction.

THE AIM OF THE BOOK

The aim of this book is threefold:

- ❑ to give a broad overview of what is meant by risk and the way in which it influences decisions made in the construction industry;
- ❑ to describe some of the tools and techniques used in risk management in a broad range of industries;
- ❑ to describe systems and techniques that could be used by the design and construction team in the management of risk on construction projects.

Chapters 1 and 2 give the background to risk and uncertainty and deal with some of the theoretical aspects of risk. Chapter 3 describes a framework for a risk management system. Chapters 4 and 5 look at the tools and techniques and the later chapters consider the application of risk management.

It is hoped that the readership will include, clients, architects, surveyors, engineers, contractors and other professionals; hence when referring to a person making a decision, the term' decision-maker' has been used.

1

PUTTING RISK INTO PERSPECTIVE

INTRODUCTION

Risk! Construction projects have an abundance of it, contractors cope with it and owners pay for it. The construction industry is subject to more risk and uncertainty than many other industries. The process of taking a project from initial investment appraisal to completion and into use is complex, generally bespoke, and entails time-consuming design and production processes. It requires a multitude of people with different skills and interests and the co-ordination of a wide range of disparate, yet interrelated, activities. Such complexity moreover, is compounded by many external, uncontrollable factors.

In view of the inherent risks in construction, it is surprising that the managerial techniques used to identify, analyse and respond to risk have been applied in the industry only during the last decade. Most people would agree that risk plays a crucial role in business decision-making: the risk of loss tempers the pursuit of return. There is less agreement about what constitutes risk. It is well-publicised and much talked about, and yet intangible. Risk can manifest itself in numerous ways, varying over time and across activities. Essentially, it stems from uncertainty, which in turn is caused by a lack of information.

Numerous texts are available which deal with the underlying theoretical concepts of risk and with techniques which identify and manage it. There is a gap between the theory and the techniques proposed to manage risk, and what people do in practice. Intuition, expert skill, and judgement will always influence decision-making, but a set of tools is now needed which will enable risk management techniques to be put into practice in the construction industry. This book is intended to be a first step in this direction.

RISK AND REWARD GO HAND IN HAND

Most people, asked to name a situation which involves risk, would perhaps think first of physically dangerous sports, such as sky diving or motor racing. Others might cite gambling, whether in poker games or the stock market. Behavioural scientists would also include risk-taking enterprises which the public would not readily identify as such, for example, getting married.

Thus, the concept of risk can be applied to nearly every human decision-making action of which the consequences are uncertain. This uncertainty arises because an essential characteristic of decision-making is its orientation towards the future - a future which by its very nature is uncertain. Time is therefore a central variable to be considered when dealing with risk. We can take risks or we can be at risk. We can speak of *the* or *a* risk and we can consider ourselves as risking something.

Risk

The word risk is quite modern, it entered the English language in the mid 17th century, coming from the French word *risqué*. In the second quarter of the 18th century the Anglicised spelling began to appear in insurance transactions.

In a manufacturing or commercial context, risk is endemic to all investment decisions. Each investor, faced with investments characterised by very different risk/return profiles, will have an individual attitude to risk. At one extreme, the investor can opt for a relatively risk free investment by purchasing government short term treasury bills issued at a fixed rate of return. At the other extreme the investor can decide upon ordinary shares; the high risk involved was demonstrated in the October 1987 (known as Black Monday) share market crash around the world.

Obviously some decisions are more important than others. Take for example the individual faced with two decisions: whether or not to take an umbrella to work and whether to invest millions of pounds in developing and building a hotel. The process of decision-making can be intuitive, pragmatic or dogmatic, or it can be rational and scientific, depending on the importance of the consequence. If he fails to take an umbrella and it rains, he will get wet and might catch a cold; if he fails to get a sufficient return from his hotel investment, the result is financial disaster and possible ruin.

Investors in financial markets recognise that risk plays an important role in their allocation of assets across different investments. In a climate of economic uncertainty, with increasing volatility in the financial markets and on the foreign exchanges, investment decisions by management have become more demanding. Estimates of expected profits or returns, which are based upon the percentage return on investment or the internal rate of return, will not provide the company or the investor with sufficient information on which to base a sound decision. The investor needs some indication of the possible deviations from the expected profit or returns if there is a downturn in the economic conditions, and of the sensitivity of the investment to changes in the market. Simply put, the investor needs to know his or her risk exposure.

The two most important questions are whether the returns on the project justify the risks, and the extent of the loss if everything goes wrong. Clearly, the decision-maker's perception of risk is more likely to be influenced by the probability of a loss and the amount of that loss than by a variance in the gamble. Thus the techniques for quantifying risk as an aid to decision-making have become more important. These techniques must be based on a proper understanding, both of the terms involved and of other basic concepts such as why, given exactly the same situation and information concerning a proposal, two people may come to different decisions.

Risk and construction

It might be argued that these considerations apply to investment in financial markets but have little to do with the apparently more 'real' environment of the construction industry. Nothing could be further from the truth. The individuals involved in the industry form two groups: 'principals' who commission construction and 'agents' who undertake the various activities that produce buildings, roads, bridges etc. These groups are, of course, heterogeneous. A principal can be anyone from a government department or a major development company to an individual householder. Agents include professionals such as architects, engineers, surveyors, general contractors, and a wide range of specialist sub-contractors and suppliers.

It is easy for the principals to see the relevance of risk management. A principal, in using the construction industry, is making an investment decision: the decision to commission a prestigious office building or a new garage. The capital committed could, instead, be invested in government bonds or some market portfolio of financial assets. The decision to invest in a building must, therefore, provide a risk/return profile which is competitive with the best that the financial markets can provide.

For the agents, the argument is not so straightforward but is equally valid. An agent bidding for the relevant part of a building project is committing resources - labour and capital - that have other potential uses. Money may have to be borrowed, or reserves used, to cover a gap between income and expenditure, while profit, if it is made, will arise at some time in the future. With regard to the agent's own financial resources being used, the same considerations arise as those outlined above for principals in that the agent could invest his resources in financial markets instead. Where the agent is borrowing or committing tangible resources such as labour, a comparison must be made between potential return and the cost of borrowing, and/or between potential return on this project and potential return on the projects that could otherwise be undertaken. Once again, effective risk management requires comparison of risk/return profiles.

Risk - another four letter word

To many organisations risk is a four letter word and they try to insulate themselves from risk. They position themselves to unload unexpected costs onto others. Increased costs are passed on to consumers by raising prices. Sub-contractors are played off against each other to gain the cheapest price.

This approach may well have worked in the past, but it can be a recipe for disaster. As international competition becomes stronger, saddling consumers with increased costs becomes more difficult. Shifting financial risk onto sub-contractors, the group least able to resist, does not encourage high levels of trust and commitment. We are moving to an era where risk has to be identified, analysed and apportioned much more openly and professionally.

Despite the apparently obvious nature of these remarks, it remains the case that most risk identification and appraisal carried out in the property and construction industry is poor in comparison with the quality of analysis used in, say, the money market. This cannot continue. Increased integration between financial and real sectors of the economy, and major capital commitments in the building industry, means that the poor quality of risk management in construction has perhaps a greater significance at present than at any other time since the 1970s.

Pundits have argued that there are four ways to tackle risk in the construction industry:

❑ *'the umbrella approach'* where you must allow for every possible eventuality by adding a large risk premium to the price;
❑ *'the ostrich approach'* where you bury your head in the sand and assume everything will be alright, that somehow you will muddle through;
❑ *'the intuitive approach'* that says don't trust all the fancy analysis, trust your intuition and gut feel;
❑ *'the brute force approach'* that focuses on the uncontrollable risk and says we can force things to be controlled, which of course they cannot.

Few companies today can survive on these approaches, clients are demanding more and everybody is becoming more conscious of the risk they are carrying.

Risk management aims to ensure that all that can be done, will be done to ensure the project objectives are achieved. Once a risk is identified and defined, it ceases to be a risk, it becomes a management problem.

AGAP (ALL GOES ACCORDING TO PLAN) AND WHIF (WHAT HAPPENS IF)

In the construction industry the euphoria, optimism and excitement of a new project often leads to the AGAP attitude. We tend to give budgets, estimates, and completion dates based upon all going according to plan. Construction has many unknowns and things rarely go according to plan.

We need to be more aware of WHIF (what happens if) analysis. People should be encouraged to have 'brainstorms of destructive thinking' where wild ideas can be thrown up about things which might go wrong, even though there is no precedent. For instance, neither Boeing or General Electric engineers had foreseen a washer breaking loose causing the Lauda Air 767 engine to go into reverse thrust, resulting in the death of 267 people. It might have been a 50 million to one chance but the consequences of the loss were catastrophic.

The ideas need to be collected into a risk management system where analysis can be undertaken.

THE PEOPLE, THE PROCESS AND THE RISKS

Any construction project involves a variety of organisations and a larger number of people (see **figure 1.1**). This is detailed further in **figure 1.2** which divides the land/building conversion into two aspects: the process and the people. The two groups that are hardest to specify exactly are the client and the user of the building, primarily because the terms 'client' and 'user' are all-embracing, unlike 'architect' or 'surveyor', which represent professional skills. Even the terminology is not uniform. In some contracts the client is referred to as the owner, or the employer.

Clients are not homogeneous: a point which will be discussed in more detail later. A client might be an owner-occupier who wants to use the building to manufacture goods, or a pension fund which wants the building as a long term investment, or a developer who has obtained finance for the project from a consortium of Japanese and American banks and who intends to find tenants for the building. The permutations are endless, but what all clients have in common is their exposure to risk.

Figure 1.1   A simplified view of the people involved in the traditional approach to contracting for a commercial building

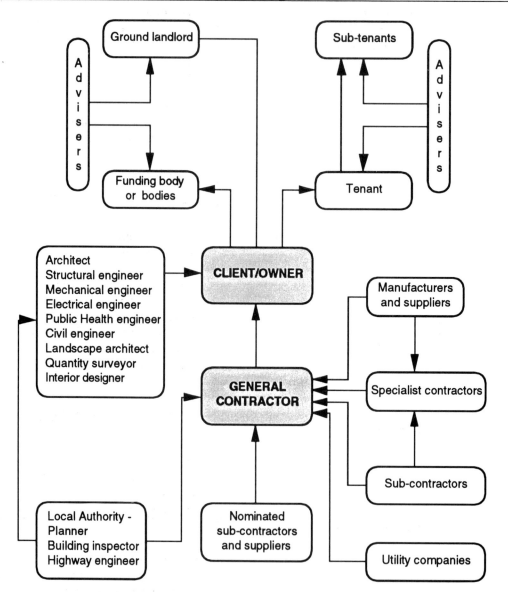

Figure 1.2   The process and the people

The Process	The People
Investment decision *(Appraisal/Feasibility/Budget)*	Clients/Owners Funding bodies Banks, insurance companies, pension funds Users Advisers - General practice surveyor (Investment/agency) - Quantity surveyor - Designers - Lawyers
Design	Project manager Architect/Planner Structural engineer Quantity surveyor Civil engineer Mechanical services engineer Electrical engineer Public health engineer Chemical engineer Landscape architect Interior designer Other specialist consultants Building inspector for building code/regulations approval
Construction	Contractor Design team Specialist contractors Sub-contractors Suppliers Manufacturers
Occupancy/use	Owner User Advisers - General practice surveyor - (Valuation/agency/management) - Building surveyor/engineer - Maintenance manager

Typical risks on a construction project include:

- ❑ failure to complete within the stipulated design and construction time;
- ❑ failure to obtain the expected outline planning, detailed planning or building code/regulation approvals within the time allowed in the design programme;
- ❑ unforeseen adverse ground conditions delaying the project;
- ❑ exceptionally inclement weather delaying the project;
- ❑ strike by the labour force;
- ❑ unexpected price rises for labour and materials;
- ❑ failure to let to a tenant upon completion;
- ❑ an accident to an operative on site causing physical injury;
- ❑ latent defects occurring in the structure through poor workmanship;
- ❑ *force majeure* (flood, earthquake etc.);
- ❑ a claim from the contractor for loss and expense caused by the late production of design details by the design team;
- ❑ failure to complete the project within the client's budget allowance.

It is important to distinguish the sources of risk from their effects. Ultimately, all risk encountered on a project is related to one or more of the following:

- ❑ failure to keep within the cost budget/forecast/estimate/tender;
- ❑ failure to keep within the time stipulated for the approvals, design, construction and occupancy;
- ❑ failure to meet the required technical standards for quality, function, fitness for purpose, safety and environment preservation.

In most situations, the effect of adverse events will be financial loss. The task of the professional advisers, contractors, specialist contractors and suppliers is to identify the discrete sources of risk which cause failure to occur, and to develop a risk management strategy that provides for the most appropriate organisation to carry that risk.

Risk and uncertainty do not only occur on major projects. Whilst size is an important consideration, factors such as location, complexity, buildability, and type of building can all contribute to the risk. It is no accident that a complex, highly serviced £5 million hospital operating theatre is going to carry more risk of a construction cost and time overrun than a £5 million warehouse building.

Furthermore, it is rare for two construction projects to be exactly alike. By their nature they are different, which means every project has to be considered afresh. Fortunately however, an effective risk management system contains a set of techniques that can be applied to any project.

CLIENTS OF THE INDUSTRY

The clients of the industry ultimately pay the bill and it is important to understand their needs and expectations.

Figure 1.3 shows the range of clients of the construction industry. Historically, clients have been separated into two groups, the public and private sectors, as determined by the source of funding for the project. In economic terms, client demand is made up of four types:

❑ *a means to further production of goods and services,* e.g. factories, offices, chemical plant, gas platforms
❑ *an addition to, or improvement of, the infrastructure of the economy,* e.g. power stations, pumping stations, sewerage plant
❑ *a social investment,* e.g. hospitals, schools, churches
❑ *an investment good for direct utility,* e.g. housing.

The private sector clients can be further divided into three groups:

❑ *the ownership group,* which wants to use the constructed facility as a factor of production, for a social function, or for housing
❑ *the investment group,* which is seeking a sound investment showing capital growth and earning revenue
❑ *the property dealing group,* which deals in property either in the short or long term; the skills of this group reside in building new or modifying existing property.

The priorities and experience of the client groups will be very different, but the attitude of the owner-occupier who is extending his home will not differ greatly from that of the developer undertaking a 50,000 sq m. floor area speculative office block. The scale of the operation of the former is smaller, but above all they both want certainty that the project will be finished on budget and on time. They don't want surprises and they do want to minimise their risk exposure to additional cost and time overruns.

The main characteristic of individuals in the ownership group is that they do not see buildings as an end in themselves but simply as a means to an end. The factory owner wants to use the building to house his expensive equipment: The quicker the building is in use the quicker the investment starts to show a return. The householder wishes to occupy his new possession as quickly as possible in order to enjoy the amenity it will provide.

Figure 1.3   The clients of the construction industry (based on funding)

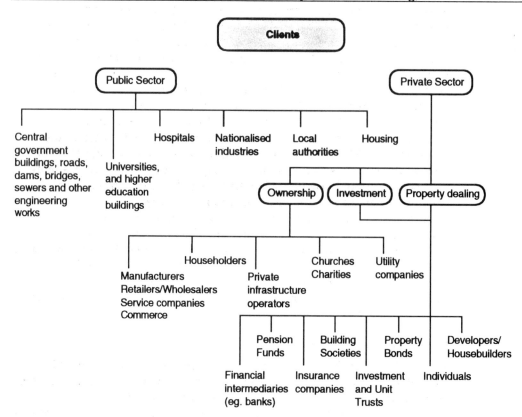

The investment group deals in buildings for investment gain. Its members do not want to use the property for production or for social need, but rather for investment growth and yield. Since the 1960s, property as an asset has become an increasingly popular medium for investment in its own right. The number and type of organisations indulging in investment has increased. Institutions such as pension funds, life insurance companies, unit trusts and managed pension funds will all invest in property. They have moved from merely owning equities, including equities in property companies, and granting loans for property development, to becoming owners of vast estates of commercial property. In general, the institutional investor is interested in constructing a total portfolio consisting of gilts, property, and equity shares. Investors are constantly seeking to adjust their overall risk position by balancing the amount of funds invested in property with other forms of investment, such as ordinary shares.

Some investors have vast investments in property. For example, Mitsui Fudosan, a Japanese conglomerate, owns 2.7 million sq m. of office space in Japan, while Mitsubishi Jisho owns 2.6 million sq m.

Individuals in the property dealing group make their living by acting as developers and speculators; they are actively involved in the acquisition of the land, the design process and the construction process. Their skills and experience are in project assembly. Property companies may or may not have their shares publicly quoted, but strictly speaking they are not institutions as they are not vehicles for saving money which is then invested or loaned. They might take an equity interest in a development or they might act as project manager for an investing institution, depending upon the risks and the potential return.

The priorities of the public sector are very similar to those of the private sector. We live in an age of public accountability and value for money; there is no bottomless pit of money to call upon. Recently, a more commercial outlook has prevailed, and the public sector has become prepared to innovate with design and build, management fee arrangements, and incentive payments based upon performance.

The public sector is a very important client, and therefore has influenced the buying of construction work. The proliferation of wordy and detailed contractual clauses, and the mass of paper that is needed to establish and confirm the contractual position, is evidence of this. The main purpose of such detail is to ensure that the bulk of the risk resides with the contractor.

Have clients' needs changed?

The desire for uniformity, standardisation and comprehensiveness has created standard forms of contract designed to cope with most conceivable eventualities, and rightly so. But economic conditions have changed rapidly over the past twenty years, and the funding of projects has become more complex, particularly in the private sector. The conflict mentality between contractor and client in the 1960s and 1970s left the industry with contracts that are very detailed and very flexible - perhaps too flexible. They allow design to be incomplete at the tender stage, and they allow clients to change their minds during the construction phase.

The complex funding arrangements involving high interest charges and risk premiums are less flexible because of the number of parties and advisers that are involved. The desire for certainty in both private and public sectors is becoming paramount, and this is coupled with a need to minimise the exposure to risk. All this has to be tempered with reality because there will always be an inherent conflict between certainty and flexibility. Clients cannot have it both ways. If they want flexibility they will have to accept some of the risk.

Privately financed infrastructure projects

New types of projects are emerging with the change in circumstances. All around the world privately financed infrastructure projects are beginning to appear, for instance, the US$435 million Hong Kong Eastern Harbour

Tunnel, the US$1.1 billion Bangkok Expressway in Thailand and the US$1 billion E-470 toll road in Denver, Colorado. The biggest project of this type is the Channel Tunnel between Britain and France.

Private sector infrastructure project (Build-operate-transfer)

Case Study: The Sydney Harbour Tunnel in Australia
In 1986, a Joint Venture company, Transfield-Kumagai (Transfield is Australian and Kumagai a Japanese company) put to the New South Wales Government a plan combining two major innovations.
The first was to link existing approach roads on both sides of the harbour with a tunnel. The second was private enterprise funding, with the Joint Venture bearing the construction and financial risks of a fixed price contract.
Late in 1986 the Joint Venture released a multi-million dollar feasibility study and environmental impact statement. The proposal was approved in 1987, construction began in 1988 and was completed in 1992.
The Joint Venture carried all the risk of cost and time overruns. That commitment was backed by a A$40 million performance bond and a A$40 million loan. The Government provided a A$223 million loan, repayable in 2022, based on projected bridge toll revenue. The remaining construction finance comes from bonds underwritten by a bank and repaid over 30 years from the tunnel tolls.
Total construction cost was a fixed price: A$486 million at 1986 values, or A$554 million at current prices over the construction period. Interest and principal repayments bring the total to A$750 million.
The tunnel will be operated privately until 2022 when, debt-free and worth an estimated A$2 billion, it will be handed over to the Government at no charge.
In operation, the Government provides an ensured revenue stream to repay bondholders and meet operating charges and maintenance costs. The revenue stream is based on conservative forecasts for harbour crossing traffic.
The Joint Venture spent considerable time and effort to identify the risks and to minimise their impact. For example, environmental considerations, noise, heritage considerations, visual impact, air and water quality, as well as the design and construction risks.

These types of projects are based upon build-operate-transfer (BOT) arrangements, where the contractor designs, builds and operates the project over a number of years. Other acronyms are BOOT (build-own-operate-transfer) BOLT (build-operate-lease-transfer) and BOO (build-own-operate). The private sector participation in infrastructure projects can result in a combination of lower capital costs and less risk for the public sector.

The projects rely on good estimates of the construction cost, the traffic volumes, the operating costs and the tolls likely to be collected over the time horizon of the lease. The projects are long term with low cash flows in the early years. Governments are unable or unwilling to commit public money to the schemes, so the domestic capital markets finance these huge projects. The investors are taking risks which are identified and analysed in considerable detail.

Many pundits argue that performance and risk are increasingly being measured in the short term. Clients want to see tangible returns quickly. However, projects which involve infrastructure development cannot be viewed in the short term and time horizons of up to 30 years have to be used.

BOT projects require very complicated risk allocation and sharing arrangements among several parties. The parties have to be satisfied that their risks are sufficiently limited, while at the same time all the project risks have to be covered to the satisfaction of the project's creditors and investors.

What do clients want?

Clients can have very different objectives, but their needs can be grouped under the headings of **time, cost,** and **quality. Time** can mean both the need for rapid construction and completion on the stipulated date. **Cost** means obtaining value for money and completing the project within budget. **Quality** is used to cover technical standards, including such areas as safety and fitness for purpose. The relative importance of time, cost and quality will vary from client to client (and between similar clients in different countries). What is, however, certain is that the clients of the industry do not want surprises. They want to achieve their desired objectives and to this end a professional approach to risk management is required.

INVESTMENT IN PROPERTY

Real estate investment is more risky that many other forms of investment. If one were to invest in, say, short term treasury bills, there would be little risk of default since the Government issues them. Such bills are priced at an annual yield, with no possibility of receiving less than that yield. Returns are guaranteed; hence the standard deviation of these returns is zero and such an investment is 'risk free' apart from the effects of inflation. However, the rate of return on bonds is much lower than on equities. Investing in equities on the stock market can substantially increase the returns but also, in recent years, there has been a considerable increase in the risk of such investment.

Property, however, belongs to a risk category which does not provide a guarantee of return. Furthermore, property, unlike equities and gilts, is not highly liquid and marketable.

Any investor seeks to reduce risk exposure by spreading funds across a range of investments. To illustrate this **table 1 A** shows a typical asset allocation of a UK pension fund.

Table 1 A   Typical asset allocation of a UK Pension Fund

Investment sector	Percent
UK equities	51
Overseas equities	20
UK fixed interest bonds	12
UK index linked bonds	3
UK property	9
Cash/other investment	4
Overseas property	1
	100

Only 9% of total investment is in property. This level of investment is reflected in the level of performance. More importantly, **table 1 B** shows how the level of annual investment in property by the pension funds and investment companies as a percentage of the asset distribution has reduced over a ten year period from 21% in 1981 to 7% in 1991. The graph is derived from a survey by the WM Company which takes account of £241 billion of investment. The decline in the investment in property ultimately affects the volume of output for the construction industry because of the work in new build and modernisation and refurbishment. **Table 1 B** shows the annual returns for individual investment sectors between 1982 and 1991. The equities market showed a good return, but the risks are high, a feature that became particularly apparent on 'Black Monday' in October 1987.

The fundamental difference between the property market and the equity and fixed interest markets lies in the individual characteristics of property investment. All properties are different, and the value of a property is the product of its location, design, age and a host of other factors making monitoring much more difficult compared to the range of performance indicators available in the stock market. Stocks are liquid, property is not, it takes a long time to complete a property transaction.

Table 1 B   Annual rates of return 1983-1991

	1983	1984	1985	1986	1987	1988	1989	1990	1991
UK equities	28.4	29.8	19.8	25.9	7.1	10.4	36.0	-9.8	20.0
Overseas equities	40.5	21.1	10.8	37.2	-18.5	23.4	40.3	-27.3	20.9
UK bonds	16.2	10.4	12.6	12.5	16.4	8.2	7.4	7.9	18.2
Overseas bonds	-	-	-	26.0	-1.0	4.6	17.4	-1.4	21.0
UK property	8.7	12.3	3.6	7.3	19.4	32.8	18.2	-7.5	-1.8
Overseas property	-	-	-	5.8	-14.1	10.7	24.2	-12.7	-0.2
Cash/other	9.3	1.5	19.9	11.6	8.1	9.2	14.2	12.6	10.4

Source: WM Company - Investment Performance Results

The above discussion is concerned with historical performance. However, when making a decision about the future, the investor is inevitably concerned with events in the future. At any point of time, the market's target return from property will reflect the investor's perception of future risk which will be influenced both by historic performance and by expectations for future economic conditions.

In general, investment in property will involve cost, revenue and value risks which result in profit or loss. Investments are concerned with:

❑ annual yield;
❑ capital value;
❑ growth in capital value over both the short and the long term;
❑ total return;
❑ gross and net rental income;
❑ growth in rental income;
❑ outgoings.

Companies investing in property will face a 'sector risk' which relates to the value of investments in a particular property sector, such as shops, offices or industrial property. For instance, when considering a shopping investment, the options would be:

❑ purchase of an existing development or undertake a new development;
❑ which location;
❑ primary or secondary site;
❑ what tenant mix;

❑ freehold or leasehold;
❑ expected minimum acceptable return;
❑ growth potential.

Increasingly, investment in property is being undertaken on a global basis. For example, there have been international investment flows from Japan and the USA into European property. Also, the UK has been a big investor in US and Australian property. Japan, Taiwan and Hong Kong are particularly important in the Pacific Rim markets.

Currency risk is a major consideration for all international investors. The Yen fell about 20% in value during 1990 against sterling. Similarly, the US dollar has dropped against European currencies. The international investor has to consider the exchange risk when making any investment decision.

Figure 1.4 The cash flow sequence for a project financed with debt finance

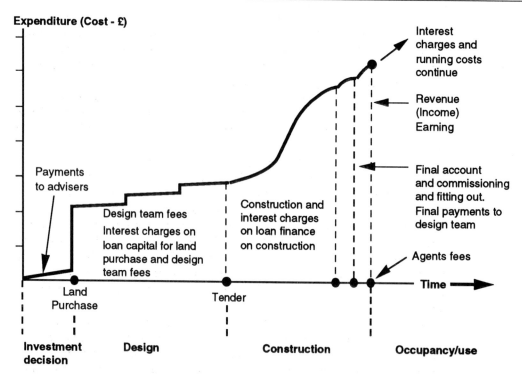

Figure 1.4 shows the cash flow for a project from the investment decision stage through to occupancy. Throughout the whole process, the client is facing expenditure, with no revenue income until the building is

occupied. The saving feature is that the site and building will have a residual value to cover the expenditure. The interest charges will relate to the amount of loan capital in the project.

The client is balancing the expenditure and income for the project, just as the contractor and the specialist contractors will be seeking to ensure that their income is above their expenditure.

CONSULTANTS AND RISK

The advisers who offer professional services to the client on the investment, the design, cost, contractual arrangements and all the other facets of dealing in buildings and property, will always seek to balance the risks in the best interests of the client.

Such consultants must use their skill, knowledge and experience with care to ensure that the client's interests are protected. At the same time, the contractors, suppliers and specialist contractors must ensure that their interests are protected. In general terms, the greater the risk a party must carry, the greater the reward that will be sought.

Consultants have an expressed and an implied professional duty of care to their clients, who, in recent times, have shown a willingness to sue for damages when they have suffered a loss as a result of bad advice. Where a consultant gives incorrect advice, they may be sued for damages under the tort of negligence. A plaintiff must of course, prove that the defendant owed them a duty of care, that this duty was broken and that damage was suffered as a result. The definition of skill and care is held to be 'such high standard as judged by practice and knowledge to be generally obtained at the date that the duties are performed'.

Lawsuits against the professionals have increased by 25% over the past five years and professional indemnity insurance premiums have doubled. Architects and building surveyors have been sued for latent defects discovered in buildings, investment surveyors have been sued for bad investment advice, and quantity surveyors have been involved over certification of interim payments on building projects in which there has been a dispute about the quality of the work. Therefore, whilst consultants can only accept the risk for their professional integrity and competence, they must manage their own risks as well as looking after the interests of the client.

One argument often used is that consultants carry very little risk because they only give advice, but the loss that arises as a result of bad advice can be very significant. For instance, in the high technology buildings located in the City of London, losses for the client could be many orders of magnitude greater than the value of the work that might give rise to the breach or act of negligence.

CONTRACTING AND RISK

The primary burden of risk on a construction project falls between the contractor and the client, and insurers will often carry low probability, high impact risks, such as fire or collapse. The risk is best placed with that party involved in the management of a project who is best able to manage the factor which gives rise to it. For instance, the risk of low construction productivity on site is a controllable and acceptable risk for the general contractor to manage.

On the other hand, factors such as inflation are beyond the control of the contractor. In times of low inflation, contractors will carry the risk, but an allowance for inflation will be included in the contractor's tender price plus an appropriate risk premium, which is an allowance for risk. Alternatively, in times of high inflation and with a fixed price tender, the contractor must allow a substantial risk premium to compensate for the high risk exposure that must be faced.

There are various contractual arrangements which apportion the risk equitably; these are discussed later. However, over the past decade, various new agency approaches have been developed, such as management contracting and construction management, which have changed the traditional relationship between the contractor and the client. In essence, the contractor acts as an agent on behalf of the client, being paid a fee for his management and organisational services. This has increased the burden of identifying where responsibility lies when something goes wrong.

For instance, in building a typical £2 million old people's home, there could be fifty different sub-contract packages. In any one of the sub-contracts, such as the mechanical and electrical services installation, there could be another four or five specialist sub-sub-contractors. In every situation, there will be a contract agreement and conditions of some type, but a small specialist sub-contractor with a small package of work might be expected to carry the potential risk of a very large claim for loss and expense if the project were delayed as a result of his bad workmanship or late delivery.

When things go wrong everybody will find good reasons why it is somebody else's fault - that is human nature. Hunting the guilty after the event is of very little help. Everybody needs to be made more aware of the risks and responsibilities from the outset. It is pointless imposing financial burdens on a company that cannot afford to pay.

When contractors act as agents they take on very different responsibilities and risks.

Under the law relating to agency, an agent must obey instructions, must not delegate his functions to anyone else without his principal's permission, and must show due skill and care in carrying out his instructions. The agents have a duty of care and they can be sued for negligence if they fail in their duty. It is not only the client who can bring an action. In a recent case in the USA, a concrete frame specialist contractor successfully brought an action against a construction manager claiming that he negligently planned and scheduled the contract works which had caused him direct loss and expense on his work.

The apparently straightforward approach of using a contractor as an agent has therefore become very complex. Problems can and do occur on projects, and it is therefore important to have a systematic approach to risk management.

2

THE BACKGROUND TO RISK AND UNCERTAINTY

INTRODUCTION

It is tempting to talk about the concept of risk at a practical level without fully explaining all the terminology and techniques. Although there is much published material on the subject, it is difficult to fit an overall picture, a framework into which all the constituent parts fit.

This chapter, therefore, looks at some of the background theory of risk and uncertainty, and begins with a discussion of the differences between them. Probability, which underlies many of the analytical approaches to risk, is defined; decision-making in the construction industry and the role of the past in forecasting are considered. The concept of utility measures an individual's attitude to risk. It is therefore as important to any discussion of risk management in the construction industry as elsewhere and this is described in chapter 3.

There have been numerous studies, both theoretical and empirical, which focus on the riskiness of situations, while others focus on the willingness of people to take risks in certain situations. The different types of risk studied include: business risk; investment risk; economic risk; social risk; political risk and so on. Risk can mean many different things, it has different definitions and can be dealt with in many ways. Merely seeing a problem from a different perspective will influence a person's perception of risk. An owner commissioning the construction of a new department store is concerned that the project is within the budget price and opens on time. The contractor has the same objective but his aim is to make a profit on the construction. His view of the risk is to manage the project within his cost estimate. Both parties are seeing the risk from a different viewpoint.

DEFINING RISK AND UNCERTAINTY

The environment within which decision-making takes place can be divided into three parts:

❑ certainty
❑ risk
❑ uncertainty

Certainty exists only when one can specify exactly what will happen during the period of time covered by the decision. This does not, of course, happen very often in the construction industry.

Many writers go to great lengths to explain the difference between risk and uncertainty, while others consider the terms to be synonymous. There is a general consensus that a decision is made under risk when a decision-maker can assess, either intuitively or rationally, the probability of a particular event occurring. Risk has its place in a calculus of probabilities and lends itself to quantitative expression. For instance, past performance indicates that a 60 storey office building can be built in the UK in 30 months; there is an element of risk in this estimated duration, but past data tell the decision-maker that it can be achieved with at least some degree of certainty. The risk of success and the cost of failure can be calculated by using the probability of failure.

Uncertainty, by contrast, might be defined as a situation in which there are no historic data or previous history relating to the situation being considered by the decision-maker. In other words it is 'one of a kind'. An example would be the building of a 60 storey office building in the People's Republic of China in 30 months. China has never built a project 60 storeys high and furthermore the decision-maker may have no knowledge of the Chinese building industry.

A company has to operate in an environment where there are many uncertainties. The aim is to identify, analyse, evaluate and operate on risks. Accordingly, the company is converting uncertainty to risk.

The more one thinks about risk and uncertainty, the more one is inclined to the view that risk is the more relevant term in the building industry. Even in the China example above, there is some information available on which to base an estimate of duration. As a result, throughout this book we confine ourselves to using the term risk to encompass uncertainty.

The uncertainty of life and construction projects

Construction projects involve hundreds or even thousands of interacting activities, each with a cost, time, quality, and sequencing problem. The costs and durations are uncertain and one response, still surprisingly common, is to shy away from uncertainty and hope for the best. Another is to apply expert judgement, experience, and gut feel to the problem. We all know of the experience of the chief estimator in a construction company going to great lengths to ensure a reasonably accurate estimate for a project, only to be told by the estimating director at the tender adjudication meeting, that 2% has to be taken off the items because of the gut feel about the market.

In spite of this, substantial investments are made on the basis of judgement alone, with little or nothing to back it up. This is the gut feel approach to risk and is discussed later.

Risk and uncertainty need to be identified in a structured way to try and remove the fog of uncertainty. Building models of uncertainty in projects is not difficult. The hardest part is not finding the techniques or the tools to analyse risk and uncertainty, but accepting that life is uncertain and that it is better to grasp it rather than ignoring it.

Dynamic and static risk

Dynamic risk is concerned with maximising opportunities; for instance it might concern developing a new and innovative product. Dynamic risk means that there will be potential gains as well as losses. Dynamic risk is risking the loss of something certain for the gain of something uncertain. Every management decision has the element of dynamic risk governed only by practical rules of risk taking.

Static risks relate only to potential losses where people are concerned with minimising losses by risk aversion.

Each day brings both dynamic risk, where loss and gain are plotted against each other, and static risk where the opportunity may simply be to consolidate and protect against losses.

A threat and a challenge

All risk involves both a threat and a challenge, where there is a fine line between a threat and a challenge. An opportunity is a threat to those who predict failure and a challenge to those who think they might win. If somebody is threat-prone they magnify potential losses and minimise potential gains.

One kind of risk taking is taking a chance on pure probabilities, the other is betting on one's skill. People are often irrational when they bet on their ability in a particular skill test. Kogan and Wallach have conducted a number of tests and shown that those who are consistently high risk takers become even higher with losses or when the odds get longer. Those who are consistently low risk takers bet low and drop out quickly when losing or when the odds lengthen. This can be seen in business with the financial collapse of major corporations, such as the Bond Corporation in Australia where many of the investments were highly speculative.

The risk takers like the challenge, but when computing the odds and risk, most people leave out one factor - the person making the judgement.

Some of the basic rules for risk taking

No matter how good or how sophisticated any analysis, it will be people who make the decision on risk. Some basic ground rules apply:

Rules for risk taking
- ❏ **Don't risk a lot for a little.**
  The Watergate fiasco was a good example of failing to follow this rule. Nobody had seriously considered the consequences of being 'found out', which eventually led to the downfall of the President.
- ❏ **Always plan ahead**
- ❏ **Always analyse both the source and the consequences of risks**
- ❏ **Devise alternative options as a contingency measure**
- ❏ **Don't use other people as an excuse for inaction**
- ❏ **Don't take risks purely for reasons of principle**
- ❏ **Don't take risks to avoid losing face**
- ❏ **Never risk more than you can afford to lose**
- ❏ **Be prepared to seek advice from the experts**
- ❏ **Consider the odds and what your experience and intuition tells you**
- ❏ **Consider the controllable and the uncontrollable parts of the risk.**

Figure 2.1   'Place your waterline low'

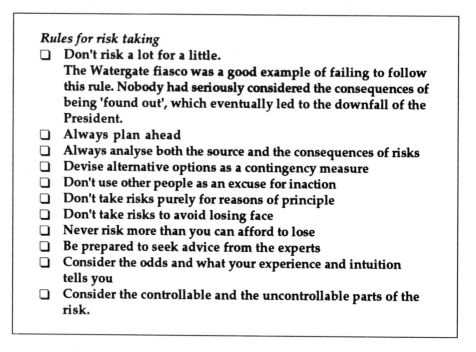

Source: Tom Peters: Thriving on Chaos

Risk - 'Place your waterline low'

Tom Peters illustrates risk by referring to the late Bill Gore (founder of W L Gore & Associates, makers of Gore-Tex and medical products). He had a superb metaphor for managing risk taking. 'You can try anything, as long as it is above the waterline.' Above the waterline meant anything that didn't affect the basic integrity of the organisation. 'If you want to drill holes below the waterline, you need to check with your boss.'

This metaphor is useful to a point. That is, the practical issue becomes **where** management places the waterline. Some companies will sit very low in the water, whereas other companies, such as financial institutions, like to sit high in the water (see **figure 2.1**).

The message is: Experiment (and risk failure) as long as the issue is trivial, never risk more than you can afford to lose. Constant risk taking and experimentation are required today simply to survive.

The risky shift phenomenon - what happens when groups make decisions

Conventional intuition is that groups, such as committees, besides being tedious, are prone to adopt very conservative policies, if indeed they don't agree to set up a sub-committee before making a decision.

However, social scientists have shown that in most cases this is not correct. When a group of people discuss a risk-taking problem, they usually arrive at a riskier solution than the average of their own previous individual solutions. Most people would disagree with this position, but research shows otherwise. The risky shift phenomenon states that groups influence decision-making towards positions of higher risk a significantly greater number of times than not, and under almost any conditions.

Social scientists are trying to determine the cause of the risky shift idea. Two possibilities exist. One is that risk taking, by implying boldness, may in society be more socially desirable than conservatism. Most people think of themselves as no less risk taking than anyone else. When opinions are aired in a group, those of lesser risk bent tend towards an increase in risk taking, seeking to be seen as courageous rather than cowardly.

A second reason is that, as a result of the emotional bond between discussants, an individual feels less of a personal responsibility for failure of risking options he would decline if deciding alone.

The risk of not risking

Perhaps the best reason for becoming a creative risk taker is summed up in a biblical quote from Mark that refers to the reward of the righteous: 'For he that hath, to him shall be given; and he that hath not, from him shall be taken away even that which he hath.' Broadly applied, this phrase reflects why risk taking is necessary. If one doesn't think big, one loses the capacity to think big. If one doesn't try the untried, soon the untried becomes frightening and mysterious.

If one lies in bed for extended periods and doesn't use one's limbs, the ability to use them at all is curtailed. If the senses aren't challenged they lose their acuity. If one doesn't take risks one loses one's capacity to do so. That is the risk of not risking.

Risk styles

There is a natural curve of risk takers, they range from those who are completely directed by others, taking all their cues from above, to those who are almost always inner directed, they make their own decisions.

Figure 2.2   People and risk

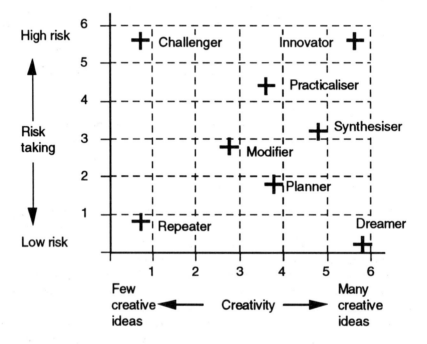

Whether people are risk taking, risk neutral or risk averse is discussed later. As a general rule, **figure 2.2** shows how people can be generalised by their creativity and risk.

We transpose this idea into people in the construction industry in **figure 2.3**. Architects are innovative and they are subjected to high risk exposure, whereas design engineers are innovative but they work to a factor of safety in accordance with stipulated design guidelines.

Figure 2.3   Construction and risk

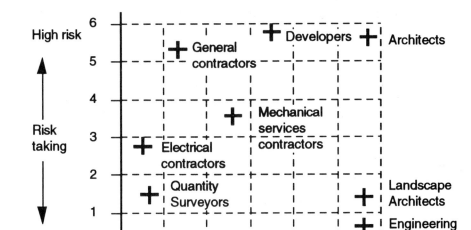

REMOVING IGNORANCE - AND RISK

The triangular envelopes below represent the amount of knowledge you need before taking any action in a given situation. We all face situations where our knowledge is inadequate. For instance, we might be a consultant civil engineer based in New York and considering how to enter the Malaysian construction market to sell our skills in waste water management. The technical press says that Malaysia needs these skills, but should we enter the market, and, if so, how?

Our knowledge tells us the size of the market and recent projects that have been awarded. We can ask local engineers about the problems of operating in the market, but we still won't know how to get the work. We will need to develop a strategy, a budget, a resource plan and a time span based upon detailed analysis.

There are four ways to remove ignorance - analysis, synthesis, simulation and test. Each one sheds light on what is meant by risk; these will be discussed in later chapters.

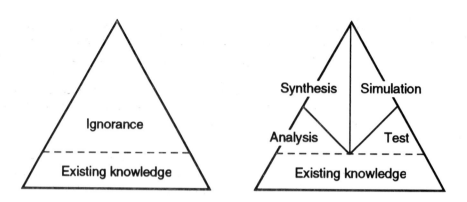

PROBABILITY

Probability is an important concept in dealing with risk, and its measurement has a long history. Definitions range from the classical deterministic notion that probability is the ratio of occurrence to the total number of equally likely cases (such as the roll of a dice or toss of a coin), to a much more subjective or judgmental view. There is no necessary contrast between the two. In some situations, such as tossing a 'fair' coin, the possible outcomes are predetermined, while in others, such as estimating the construction cost, the outcomes are much more fuzzily determined. In these latter cases, intuitive estimates may be as reliable as formal estimates; even intuitively our minds use probability to formulate our judgement.

We use probability in everyday life. For instance, weather forecasters in the United States may say there is a 30% chance of rain on a particular day. This means that for a given geographic area, on 30% of the days for which the forecast is made, it actually does rain a pre-defined minimum amount; on seven out of ten days it should not be expected to rain.

Probability theory deals with events of a special kind, called random (or stochastic) events, whose outcomes are affected by chance. In decision-making probability acts as a substitute for certainty.

There are two schools of thought about probability theory:

❑ *Objective probabilities* - The objectivists believe that probabilities must relate to long term frequencies of occurrence. In other words, only events that can be repeated over a large number of trials may be governed by probabilities. Therefore, only after repeated observations can we speak of the relative frequency of events and the associated probabilities;

❑ *Subjective probabilities* - According to this concept, the probability of an event is the degree of belief or confidence placed in its occurrence by the decision-maker on the basis of the evidence available. Hence, if the decision-maker feels an event is very unlikely to occur, he will assign a probability value of its occurrence close to zero; if he believes the event is very likely to happen he will assign a probability value close to one. Subjective probabilities represent the degree of belief that a person believes is correct, based on the information available to him. In order to use subjective probabilities in decision-making, they have to be accurate, reliable, calibrated and coherent.

In general, decisions are likely to be determined by subjective probabilities in the construction industry. Buildings tend to be unique; construction is not a factory line routine repetitive process. In most cases in the industry, decision-makers must generalise from experience and samples using relative frequencies of occurrence. All objective and subjective evidence currently available should be used in the assignment of subjective probabilities. These probabilities should reflect the decision-maker's beliefs.

Converting uncertainty to risk

Uncertain situations can be changed to risky situations by the assignment of subjective probabilities. This is not as much of an arbitrary procedure as might appear at first sight. There are several good reasons why subjective probabilities may be not only quite acceptable but even preferable to objective probabilities.

The subjective probabilities of different individuals with the same experience and information may be very different. The decision-maker's experience, education, values, personality, and perception, as well as a preference for a particular event, will be reflected in the subjective probability. Familiarity is important; when we have lived with risk we are more prepared to accept it. Personal differences between decision-makers therefore, will and do play an important role in decision-making. Subjective probabilities explicitly recognise this at an early stage in the analysis.

There remains the question, of course, of how such probabilities can be identified. If we simply asked for the probabilities from a manager, investor or entrepreneur, we might be in danger of prejudging the outcome. Although by making the decision-maker think carefully about his or her expectations, the exercise may have great value, there are procedures used to obtain numerical values of probability. One such technique is Savage's gamble method.

Let us assume that there is a choice of whether or not to open an overseas office. In order to change this situation from one of uncertainty to one of risk, the decision-maker must assign probabilities to the outcomes. One outcome is the office is successful and obtains work in its first 12 months of operation and the other is the office obtains no work; the choice is between two gambles. One is a real world gamble and the other is a hypothetical gamble for which we know the probability. So if one gamble is chosen in preference to the other, we have an indication of the subjective probability of the real world case. For example:

❏ *Real world gamble*: The overseas office obtains two major projects and earns £100,000 in fees; if there are no commissions, the direct cost of the operation of the overseas office is £50,000.
❏ *Hypothetical gamble*: A box contains ten balls, four red and six black. If you draw a black ball you receive £100,000; if you draw a red ball you must forfeit £50,000.

We know that the probability of drawing a black ball is 0.6. We also know that the payoff is the same for success or failure in both examples. So if the decision-maker chooses the real world example, we know that he or she assigns a probability of greater than 0.6 to the successful outcome of the overseas office. We can go on asking the decision-maker to choose between the real world gamble and different hypothetical gambles until we find the subjective probability, the point at which the decision-maker is indifferent between the two gambles. This point will mark the subjective probability of the operation of the overseas office. The example used above is, of course, highly simplified. Savage's method can be applied to much more complex cases with a wider diversity of possible outcomes; for example, by introducing additional colours to the box of balls.

In applying Savage's method the assumption is that decision-makers are both consistent and coherent. Consistency means that the individual assigns the greatest probability to the likeliest outcome, with the next highest probability assigned to the next likeliest outcome and so on. Coherence means that the subjective probabilities must not give rise to a situation where an unfavourable outcome is bound to occur. Negative thinking is definitely not expected!

DECISION-MAKING IN THE CONSTRUCTION INDUSTRY

There are three key factors to decision-making:
❏ **things we know**
❏ **things we think we know**
❏ **things we do not know.**

To take a risk is to take a gamble. We have to decide what to do. The more we discover about the exact nature and level of risk we face, the better we are able to prepare for it. We calculate odds, weigh up all the facts and apply experience, knowledge and guesswork. We can be confronted with risk, chance, uncertainty, probability, choice and decision. Knowledge includes both empirical data and insights obtained by interpretation. We have to distinguish between what we know and what we don't know. Mistaking the known for the unknown can be disastrous. Bankruptcy courts are filled with people who 'knew' things that were just not so. It is as important to realise what you do not know as to establish what is known.

Decisions about the source and consequences of risk are made every day, in all aspects of construction, whether it be investment decisions from the perspective of the client, design decisions from engineers or architects, or decisions concerning economy from quantity surveyors. Many of these decisions are opaque in the sense that they are hard to understand, solve or explain. They often involve numerous performance objectives. These objectives are generally conflicting, each making demands on scarce resources such as time, money or technological capability. Rarely do we consider how we have arrived at a solution, or how we justify the use of our chosen approach to decision-making.

The goal of all decision-making techniques is to map out the probabilities, consequences, and financial options, with the intention of constructing some kind of balance sheet that can provide guidance to decision-makers.

What can be done when coping with uncertain situations?
- ❑ **Ignore them**
- ❑ **Seek additional information**
- ❑ **Make more accurate forecasts**
- ❑ **Consciously adjust for bias**
- ❑ **Revise the rate of return by adding a risk premium**
- ❑ **Transfer the risk**
- ❑ **Seek alternative options.**

Decision-making is a game of imperfect information involving the future, change and human action and reaction. Most of the important decisions we make are not dealing with problems that will have right or wrong answers, but rather only better or worse solutions, hence there is a need for analysis and debate. Life can only be understood backwards but it must be lived forwards.

> *Some basic components of decision-making:*
> ❑ **The objectives of the decision-maker, ensure they are clear and simple**
> ❑ **The range of choices open**
> ❑ **The factors which must be taken into account**
> ❑ **The possible strategies that could be adopted, having due regard for coping with the uncertainties**
> ❑ **The analytical techniques to be used to aid the decision-maker**
> ❑ **The attitude to risk of the decision-maker**
> ❑ **The consideration of time preferences (short or long term) and the timing of the decision**
> ❑ **The recognition of the bias of the decision-maker and ensuring consistency.**

Decisions are rarely clear cut, we will always 'wish' we had more information on a subject. We deal with two types of problems: type A and type B.

Type A	**Type B**
simple and analysable	complex
solutions forseeable	solutions unforseeable
solutions predictable	solutions unpredictable
A leads to B	A may or may not lead to B
occurs routinely	not routine
similar to a previous project	resembles nothing done in previous projects

Type A problems rarely occur in construction, most decisions have to be made on type B problems.

INTUITION

Decision-makers rely upon both intuition and formal models to assess the worthiness of an alternative. Many decision-makers place great emphasis on intuitive reasoning, following their 'feelings' rather than their 'thoughts'. As Isenberg (1985) points out, intuitive thought is not the opposite of rational thought. It is based upon both the accumulation of experience - which allows the decision-maker to perform well learned operations rapidly - and on mental leaps which enable him to synthesise seemingly isolated information to produce results which represent more than the simple sum of parts. Intuition is the acknowledgement of some 'gut feel' about a situation and the best course of action to take. Whilst this is probably rooted in experience, it is much more tenuous and difficult to define. Decisions based on experience in some sense can be justified, whilst those based on intuition cannot.

Experience is built up over time through individuals working in and developing an understanding of, some aspect of their work. The experience can reside in an individual or in the corporate experience of the company which has been shared by the individuals within it. Nevertheless, reliance upon 'gut' feelings frequently results in poor decision-making.

Experience, intuition, judgement and 'gut feel' have their rightful place in decision-making; often they present the only legitimate and available recourse. However, relying solely on experience, professional feel and intuitional hunches does not guarantee that a best course of action will be chosen. We may solve the right problem and implement the solution skilfully, but the solution itself may be inferior or wrong.

Table 2 A   Bias and its effects

Bias	Effects
Availability	Judgements of probability of easily recalled events are distorted
Selective perception	Expectations may bias observations of variables relevant to a strategy
Illusory correlation	Encourages the belief that unrelated variables are correlated
Conservatism	Failure to sufficiently revise forecasts based on new information
Law of small numbers	An over estimation of the degree to which small samples are representative of population
Wishful thinking	The probability of desired outcomes judged to be inappropriately high
Illusion of control	Over estimation of the personal control over outcomes
Logical construction	'Logical' construction of events which cannot be accurately recalled
Hindsight bias	Over estimation of the predictability of past events

Bias and intuition

Good decisions are based upon sound analysis and intuition. Facts help to formulate the basis for the decision and intuition guides us to the decision - there has to be a balance between analysis and intuition, on occasions, you also need some luck.

The judgmental ability of humans is flawed by numerous biases, which distort the perception of reality. These biases affect the way we interpret the past, predict the future, and make choices in the present. **Table 2 A** lists the main biases which have been identified by psychologists.

These biases or heuristics (rules of thumb), contribute to what is described as the decision-maker's cognitive structure, which dictates the way things are perceived. Two common biases worthy of note are those of 'availability' and the 'illusion of control'. The availability bias is the tendency of a decision-maker to judge a future event as being likely if he can easily recall past occurrences of the event. This may indeed be a good measure of probability, since frequently occurring events are more readily recalled; but recall tends to be biased in favour of recent events, and those which appear as dramatic occurrences. For example, we recall, and are likely to assign a high probability of recurrence to 'disasters' vividly described by the press. Such occurrences are, in reality, relatively rare, and in no way compare to the dangers associated with, for example, car travel, which are rarely as newsworthy.

The illusion of control describes the tendency of decision-makers to overestimate their skill or the impact it will have on the outcome. This results in a tendency to express an expectation of success which exceeds the objective probability. This form of 'wishful' thinking can have dramatic repercussions in construction, when a commitment to a cause such as technological innovation or energy saving can blind decision-makers to the inherent risks.

Formal models clearly have a role to play in revealing the blind spots in intuitive reasoning, particularly when the complexity of the decision makes it opaque. We are less often able to base decisions on past experience because of the uniqueness of many modern construction problems which demand a more analytical approach.

Experts and experience

An expert is an individual who has some degree of training, experience and/or knowledge greater than that in the general population. In general, they are substantive experts who, in a given domain, assess events in their field of expertise.

Experience plays a valuable role in the way an expert works. Experience serves as a data base that can be used to fill gaps in the details learned about unfamiliar circumstances. The mind searches that data base of experience almost instantaneously, on an unconscious level. Over the years our minds develop categories, methods and filing systems for all their experiences. It is this fund of previous experience that helps the brain separate the relevant from the irrelevant, all without interrupting conscious thought processes.

Experience can lead to bias in decision-making, for example, people will believe a road accident is more likely to happen to them after witnessing an accident. Also, when something happens to a person he/she takes that event as being representative, when often it is not. It is tempting to solve today's or next year's problems on the basis of extrapolating the past into the future. It takes a lot of wisdom, skill and nerve to use information that disagrees with past experience.

Experience is the strongest resource available to the decision-maker, it is also the most likely cause of a blinkered approach. People feel comfortable with information that validates their previous experience, they are reluctant to use data that is hostile or discomforting to their view.

Rules of thumb

When we make a decision we sometimes use 'rules of thumb'. We process information and then apply judgement. The use of 'rules of thumb' or heuristics is widespread in the construction industry. There are rules used from banking and country credit analysis to pricing in industry. For example, international bankers lend up to two and a half times a country's foreign earnings. These rules provide guidelines for managers in decision-making. They enable decisions to be made more quickly. This is how humans evaluate information from eyesight; it is how the brain works.

The use of heuristics has become evident in the widespread use of expert systems. For sub-contractors there may be a set of criteria which potential projects must pass before the sub-contractor will take on the job. These could be rules of thumb such that the size of the job must not exceed some proportion of the firm's turnover. The job may have to satisfy some minimum criterion such as, it must not take up more than 20% of the labour force.

Making a model

Models help us to reduce our reliance upon raw judgement and intuition. The inputs to the model are provided by humans, but the brain is given a system on which to operate. Models provide a backup for our unreliable intuition.

A model can be thought as having two roles:

❑ It produces an answer
❑ It acts as a vehicle for communication, alerting us to factors we might not otherwise consider.

There is a fog of uncertainty and models give us a mechanism by which risks can be communicated through the system. A risk management system is a model, it provides a means for us to identify, to classify, to analyse, and then respond to risk.

Reacting to information

This model has its origins in the psychology of perception and memory. When a human is presented with a stimulus, a certain amount of information enters the brain. While this information is stored, a process of pattern recognition occurs, involving an interaction between the stored information and previously acquired knowledge. We know that a table is a table because we have seen it before. A second process is also taking place, this is called attention. Attention focusses on key information while filtering out the rest, so that our table is a table because it has four legs and a flat top. The way the legs are joined to the table top is ignored, unless you are a joiner.

This information can be stored in the short term memory, a memory that holds chunks of information. The sensory information is thus stored as a chunk of meaningful information, the information is interpreted. Finally, information can be stored in long term memory. This information can be retrieved in a variety of ways which suggests a complex, organised system based upon content. Further, this process appears to be a sequential one.

This information processing theory incorporates the idea of bounded rationality which holds that rational behaviour occurs within constraints. The decision-maker has a limited knowledge of possible outcomes and this knowledge is fuzzy. The decision-maker then sorts out the outcomes into acceptable and unacceptable outcomes. The sub-set of acceptable outcomes are then considered. The relation between actions and acceptable outcomes is the key to what action to undertake. Note here that the goals have become acceptable outcomes. The actions are subject to the constraints of acceptable outcomes. Further, the behavioural theory of the firm suggests that firms reach satisfactory outcomes and do not continue the search for an optimal outcome.

Most human decision-making, whether individual or organisational, is concerned with the discovery and selection of satisfactory alternatives; only in exceptional cases is it concerned with the discovery and selection of optimal alternatives. This view suggests that people simplify problems and use mental short-cuts.

Simon (1957) distinguishes between programmed and non-programmed decisions. Decisions that were repeatable and highly structured are programmed. These sorts of decisions are more amenable to computer modelling. For example, the tendering process can be said to be a programmed type of decision-making procedure.

LOOKING AT THE PAST TO FORECAST THE FUTURE

How we use the past to forecast the future is now considered.

Any look into the future involves forecasting. **Figure 2.4** illustrates a model of the forecasting process. Forecasting is a non-mechanistic process which is not restricted to a purely mathematical evaluation of trends, such as the use of regression analysis, Box-Jenkins techniques or dynamic programming. To be of use, it requires breadth of vision and experience as well as competence in employing the forecasting methodology. It relies upon elucidating significant trends from past data.

We infer from the past into the present and then into the future. We do this by means of a doctrine of 'what was in the past will continue unless something happens to change it'.

There are two stages to this process. Firstly, we infer what the future will be like before our proposed action. Secondly, we infer any effects our action might have. The future is, of course, uncertain: our inferences may be wrong. Some unexpected occurrence may invalidate our assumptions, or we may make incorrect or badly informed inferences of the future. This is the risk inherent in human actions. What forecasting tells us is that one way of dealing with such risk is to look at past experience. Indeed, just as we use past experience to infer the future, we may also use past experience of risk to infer the future riskiness of decisions.

Figure 2.4 The forecasting process

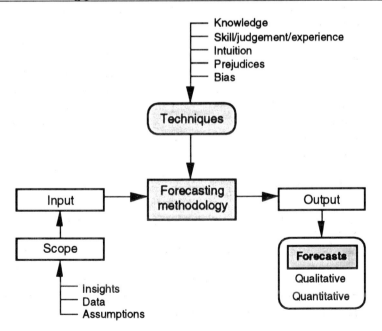

Whilst each forecasting technique has strengths and weaknesses, every situation is limited by constraints like time, funds, or data. Some important considerations are:

❑ *Data availability*

It is pointless having a highly sophisticated technique if there is a lack of available data. The extent, accuracy and representativeness of the data are important.

❑ *Variability and consistency of the data.*

❑ *Time horizon*

The time horizon of the forecast will affect the accuracy of the results. A forecast two years into the future has more chance of being correct than a forecast for 20 years. Furthermore, the time available to produce the forecast is likely to also have an impact on the accuracy and reliability of the forecast.

❑ *Cost of producing the forecast.*

❑ *Accuracy and reliability*

Risk analysis helps to identify the range of possibilities for a forecast; most importantly it allows us to ask a series of 'what if' questions. It also enables the probabilities of events happening to be examined.

❑ *Having an open mind about the future*

KEEPING AN OPEN MIND

A good example of having an open mind can be illustrated by the story of a policeman who sees a drunk scrambling around on his knees under a street lamp. 'I am looking for my keys, I lost them over there,' the drunk says pointing into the darkness. 'So why look here?' asks the policeman. 'Because this is where the light is!' replies the drunk.

Models based solely on the past (where the light is) cannot be expected to predict an ever more complex and uncertain future. Total reliance on the past can blinker the thought process and constrain lateral thinking.

TYPES OF INFORMATION

A distinction should be made between two types of information that are available to the forecaster; singular and distributional.

Singular information consists of evidence about the particular case being considered. Distributional information consists of knowledge about the distribution of outcomes in similar situations.

For instance, when forecasting the price of a new sewer installation, the design details for the sewer are known - this is singular information. What one knows about the installation costs of similar types of sewers is distributional information. The singular information describes the specific features of the design that distinguishes it from others, whilst the distributional information characterises the outcomes that have been observed elsewhere.

BUILDING A DECISION MODEL TO SOLVE A PROBLEM

The purpose of any decision model is to apply a sequence of transparent steps to provide such clarity of insight into the problem that the decision-maker will undertake the recommended action. The steps involved in such a model generally conform to those shown in **figure 2.5**.

Figure 2.5   Analysis and synthesis of problems

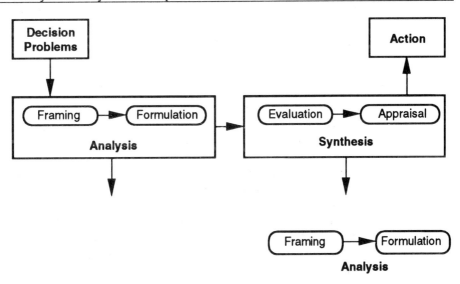

The first two stages of framing and formulation are processes of analysis, since they involve the decomposition of the problem into its constituent elements. The subsequent stages of evaluation and appraisal involve synthesis, whereby the parts are combined into a whole, to establish the worthiness of each possible solution.

Framing

The purpose of framing is to 'avoid working on the wrong problem'. A decision-maker is often unable to state the precise nature of a problem and the objectives that are to be pursued.

Far better an approximate answer to the right question which is often vague, than an expert question which can always be made precise.

(Annals of Mathematical Sciences, 1962)

Framing is what Freud described as the 'presenting problem'. He found that new patients rarely expressed their true concerns during initial conversation, and only by further probing was he able to establish their true underlying concerns. The analyst in this context might be the architect, engineer, surveyor or indeed anyone wishing to interpret the needs of the client. One problem is that analysts of different professional backgrounds often attend to different features or cues. The onus on the analyst therefore is to place a balanced emphasis on the objectives expressed by the client.

Most problems involve several objectives. To illustrate this, consider a situation where, given the means (labour, materials, finance), the designer has to construct a building with maximum energy efficiency. This involves only one objective and is clearly a technical operation. Supposing instead, the designer is expected to design the best building, as is usually the case, the concept of 'best' would then entail many attributes such as cost, risk, maintainability etc., which must be identified in the framing step.

Formulation

This step provides a formal model based upon the decision-maker's opaque problem. It is achieved by the development of a 'decision basis' composed of three parts:

❑ The alternatives available to the decision-maker to achieve the particular goal
❑ The information that describes the relationship between the decisions and possible outcomes
❑ The preferences of the decision-maker.

These are illustrated in the decision structure in **figure 2.6.**

Figure 2.6   The decision structure

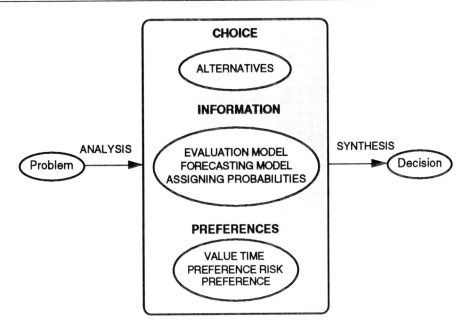

Choice
Alternatives available to the decision-maker may be readily apparent or
may need to be generated using formal tools such as brainstorming.

Information
Information includes any form of model, forecast or probability assignment
which indicates the possible outcome of the decision. Models typically
involve variables. These are either decision variables whose values are
determined by the particular alternative undertaken, or state variables
that describe environmental factors which influence the outcome, but
which are independent of the decision. For instance, any variable which
describes the likely performance characteristics, or cost attributes of an
alternative (e.g. initial cost, aesthetic appeal, energy efficiency) is a
decision variable. In contrast, variables such as electricity prices or labour
wage rates are independent of the decision undertaken. Both types of
variable are subject to risk which are often tied to a probability
assignment, describing the likelihood of occurrence.

Preferences
The decision-maker's preference must be established. Does doubling the
reliability of a component double its value to the decision-maker or is
there a diminishing value? How much more desirable are benefits
occurring now rather than later? Preferences such as these have to be

structured by the decision-maker. Formal measures such as understanding utility measures and risk preferences provide methods for capturing such preferences.

Evaluation
This stage involves the synthesis of all the data in order to establish a ranking order of the options.

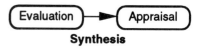

Appraisal
This final stage in the decision process examines the sensitivity of the decision and the effect of risk on the ranking order.

Table 2 B   The approach of the efficient decision-maker

Frame	Surveys the full range of objectives to be fulfilled and the values implied by choice
Alternatives	Thoroughly canvasses a wide range of alternative courses of action
Information	Carefully weights whatever is known about the costs and risks of negative consequences, as well as the positive consequences that could flow from each option
	Intensively searches for new information relevant to further evaluation of the options
Evaluation	Correctly assimilates and takes account of any expert judgement and risk exposure, even when the judgement does not support the course of action initially preferred
	Re-examines the positive and negative consequences of all known alternatives, including those originally regarded as unacceptable, before making the final choice
Implementation	Makes detailed provisions for implementing or executing the chosen course of action, with special attention to contingency plans that might be required if various known risks were to materialise

What is a good decision, as distinct from a good outcome? A good outcome is a future state of the world that we value more highly than other possibilities. By contrast, a good decision is an action that is logically consistent with the alternatives we perceive, the information we have and the preferences we feel. Good decisions do not always result in good outcomes. The presence of risk and the absence of information may alter the expected outcome. Mann (1977) has identified seven ideal procedural criteria associated with good decision-making, which are listed in **table 2 B**. The decisions which satisfy most of these criteria have a better chance of attaining the decision-makers' objectives.

3

THE RISK MANAGEMENT
SYSTEM

INTRODUCTION

Attention to risk is essential to ensure good performance, whether you are managing a company, a project or a work package. Few would deny the importance of risk management, but few analyse the risks in practice other than by using intuition and experience. Is it enough to be aware of the risks or should we try to quantify them and build mathematical models? Should we be using sophisticated computer software to provide detailed results based upon sometimes crude data?

Risk management is not new, nor does it employ black box magical techniques. It is a system which aims to identify and quantify all risks to which the business or project is exposed so that a conscious decision can be taken on how to manage the risks.

Risk management is not synonymous with insurance, nor does it embrace the management of <u>all</u> risks to which a business is exposed. In practice, the truth lies somewhere between the two extremes. A risk management system must be practical, realistic and must be cost effective. Risk management need not be complicated nor require the collection of vast amounts of data. It is a matter of common sense, analysis, judgement, intuition, experience, gut feel and a willingness to operate a disciplined approach to one of the most critical features of any business or project in which risk is generated.

The depth to which you analyse risk obviously depends upon your circumstance. Only you can judge the importance to be placed on a structured risk analysis. Conventional education does little to foster an awareness of how unpredictable reality can be. Most patterns of education concentrate on an idealised view of the world. We are taught that quantities have well defined values and that the only barrier to understanding them is our capacity to handle the mathematics.

> **Risk management is a discipline for living with the possibility that future events may cause adverse effects.**

DEVELOPING A RISK MANAGEMENT FRAMEWORK

The process of risk management is broken down into the risk management system in **figure 3.1** which shows the sequence for dealing with risk. Naturally the risk management system must be applied to each option under consideration.

Generally, the stages are:

Risk identification	Identify the source and type of risks
Risk classification	Consider the type of risk and its effect on the person or organisation
Risk analysis	Evaluate the consequences associated with the type of risk, or combination of risks, by using analytical techniques. Assess the impact of risk by using various risk measurement techniques
Risk attitude	Any decision about risk will be affected by the attitude of the person or organisation making the decision
Risk response	Consider how the risk should be managed by either transferring it to another party or retaining it.

Figure 3.1   The risk management framework

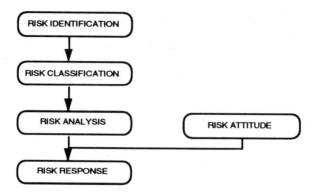

RISK IDENTIFICATION

Figure 3.2 shows the factors to be considered in the risk identification phase; the various aspects are discussed in sequence. It is worth stating that an identified risk is not a risk, it is a management problem. Inevitably, bad definition of a risk will breed further risk.

Figure 3.2   Risk identification

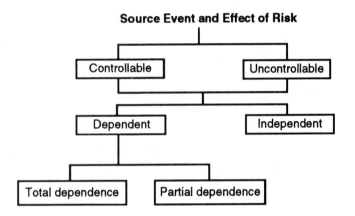

When attempting to identify risk, it is rather like trying to map the world. Maps of the world tend to be centred on the location of the map maker. Much of the world is not visible from where you stand. Some territory which is familiar and obvious to you may not be so obvious to everyone. Similarly, looking at a large project from the top, with multiple layers of planning, complex vertical and horizontal interactions, and sequencing problems, resembles looking into the world map through a fog. Management's ability to influence the outcome is limited to what they can see. The great temptation is to focus upon what should happen, rather than what could happen. A clear view of the event is the first requirement, focussing on the sources of risk and the effect of the event.

SOURCES OF RISK

The sources of riskand the effects of risk, which must be clearly distinguished, are listed below. The sequence is:

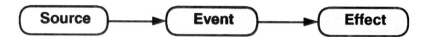

Some risks are controllable, such as a lack of co-ordination between the mechanical engineering services specialist contractor and the suspended ceiling specialist contractor. Other risks, such as the risk caused by exceptionally inclement weather, cannot be controlled, but a contingency provision can be made by considering the worst eventuality.

It is not always obvious to think in terms of the source, the event, the effect. For instance, the event of a boiler exploding might have been caused (source) by defective design, or defective workmanship resulting in the project being delayed and costing more. In the case of late completion of a project the contractor might be liable to pay liquidated damages and the client suffers the consequential loss (effect). The following list shows the approach to considering the sources, the event, and the effect of risks in a structured and systematic way.

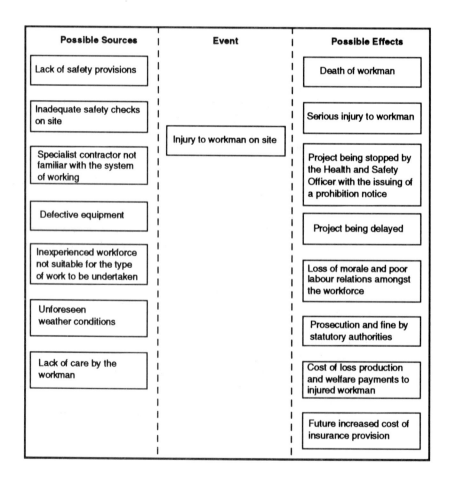

Controllable versus uncontrollable risk

There are four types of risk:
- those factors that are within your control;
- those in the control of others with whom you are forced to interact, for example the demands of the Planning Officer, Building Inspector, Fire Officer, Bank Manager;
- those that are a function of government action, such as a change in the Planning Regulations, Building Code or the rate of taxation;
- those factors that are outside of your control, for example the weather.

SOURCES AND EFFECTS

The *sources* of risk must be distinguished from the *effects* of risk

The sources of risk can be:
- inflation rising above the allowance in the estimate;
- unforeseen adverse ground conditions;
- exceptionally inclement weather;
- late delivery of crucial materials, for instance after a fire at a suppliers' works;
- incorrect design details, such as the wrong size beams being shown on the architect's drawings;
- insolvency of the main contractor;
- no co-ordination, for instance between the mechanical services contractor's drawings and the suspended ceiling specialist's drawings.

The most serious effects of risk are:
- failure to keep within the cost estimate;
- failure to achieve the required completion date;
- failure to achieve the required quality;
- failure of the project to meet the required operation needs;
- damage to the property as a result of fire or flood;
- injury to a worker due to an inadequate system of working.

Controllable risks are those risks which a decision-maker undertakes voluntarily and whose outcome is, in part, within our direct control. This contrasts with uncontrollable risks, which we cannot influence.

Driving a car is a controllable risk; a person voluntarily chooses to travel by car as opposed to other forms of transport. The increased risk of injury relative to rail travel may be tolerated because of increased convenience or comfort. Moreover, we have some control over the outcome of the journey by exercising skill and judgement.

By contrast, weather conditions, an uncontrollable risk, are entirely beyond the control of the decision-maker. Nevertheless, their adverse effect may be integrated by taking appropriate action in scheduling and in the organisation of the site.

The distinction between controllable and uncontrollable risks is important in the construction industry. In an example of controllable risks, a decision-maker might voluntarily accept the increased risks associated with new technologies in buildings. Such risks may be performance risks, where the true technical capabilities are uncertain, or financial risks, resulting from uncertain installation or support costs. These risks may be tolerated if additional benefits such as prestige, accumulation of expertise or favourable financial outcomes are likely to occur. By exploiting available expertise and through careful planning, we may be able to control the eventual outcome.

Involuntary or uncontrollable risks usually emanate from the external environment, the political, social or economic spheres. Risks associated with weather conditions, inflation, or taxation changes cannot be influenced by the decision-maker, although he can usually reduce the degree of exposure to such risks. For example, the financial consequences of increased gas prices can be reduced by the design of a more efficient plant. Indeed one could entirely 'design out' the risk by opting for an alternative fuel source.

The particular type of risk will dictate the nature of the risk response. In the case of controllable risks, the decision-maker will need to examine what measures or resources are available for ensuring a favourable outcome. He will also need to explicitly justify the voluntary risk in terms of expected benefits. Large uncontrollable risks need to be examined, to establish a means of minimising risk exposure.

Dependent and independent risks

Two sources of risk in an investment project are dependent if a knowledge of the magnitude of one, influences the estimates for the other. There may be dependency between the set of controllable risks and the set of uncontrollable risks, as in the following example.

The expected life of a building component is mainly dependent on its design, the standard of workmanship, and the quality of materials. When it is installed in the building, the component will be subject to deterioration due to ageing and physical wear and tear. It will also be subjected to abuse due to inadequate or poor maintenance. Its life will be influenced by technological obsolescence and fashion, as in the case of sanitary fittings where, despite being perfectly functional, they are replaced with the later design fittings. Most of these factors are controllable, but some, particularly those influenced by the environment, are uncontrollable.

When undertaking any risk evaluation, the question of dependence between the variables has to be considered and assumptions have to be made. There will be three types of dependence:

❑ no dependence because the variables are mutually exclusive
❑ total dependence
❑ partial dependence.

As an example of partial dependence, consider how the overall price per sq. m of the gross floor area of a 20 storey office building would be affected by an increase to 21 storeys. Trades such as floor finishing, and painting and decorating, are unlikely to be affected, but the substructure cost will increase due to the increased loads. Hence, when the variables, price per sq. m and the number of storeys, are considered, there is a partial dependence between certain of the items.

RISK CLASSIFICATION

There are three ways of classifying risk: by identifying the consequence, type and impact of risk, as shown in figure 3.3.

TYPES OF RISK

Nearly all the work classifying the types of risk is related to portfolio theory, which considers investment in stocks, equities, and gilts. Such risks are divided into market risk, which is related to the way the general market behaves, and specific risk, which is specific to a particular company.

Almost all shares respond, to some extent, to movements in the market and this response can be measured by the beta coefficient. For example, ashare with a beta coefficient of 1.5 will, on average, move 1.5% for each

Figure 3.3   Risk classification

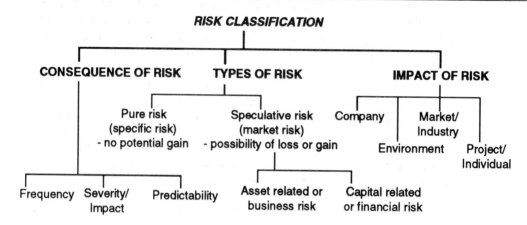

1% move in the market. A diversified portfolio of such shares would be 1.5 times as variable as the market index. The beta coefficient is thus a measure of the variability of the portfolio. Specific risk is measured by the beta coefficient for a particular company's share price against the market or sector share movement.

Portfolio theory, which is concerned with the construction of an investment portfolio, provides some useful and pertinent observations for the property development and construction industry. Like financial managers, for whom the theory was originally devised, executives from these sectors have to evaluate the economic risk of individual projects. It follows that development and construction appraisal and cash flow forecasting will involve discounting procedure to reflect individual project risk in the indeterminate future.

According to the theory, total risk is made up of market risk and specific risk. Investing in a single security implies acceptance of total risk, a move which is inadvisable since the exposure to risk is more than is necessary for the expected return. Portfolio risk can be considerably reduced with an increased number of investment holdings. Volatile events peculiar to individual companies and which are unrelated to general trends in the economy, tend to cancel out.

However, not all risk can be removed through diversification. The performance of all companies is dependent on the economy. Changes in the money supply, interest rates, exchange rates, taxation, the prices of commodities, government spending and overseas economies affect all companies in varying degrees. Holding a weighted average of all securities would eliminate specific risk. Hypothetically, an investor can hold the market portfolio which synchronises with the economy but even then, market risk remains.

One of the principles of the theory is that investors should not expect to be rewarded for taking on risks which can be avoided. Instead, they should expect to be rewarded only for unavoidable or market risk. Thus portfolio theory is in agreement with the adage of 'not keeping all your eggs in one basket', since efficient capital markets will not offer rewards for specific risk.

This argument is valid in the construction industry. Contractors, who diversify into the various sub-segments of the market such as civil engineering, oil and gas engineering and building, new and maintenance work, public and private, and who spread their activities throughout a wide geographical region, are more resilient to economic shocks. The same is true for the developer: speculative housing and offices should be complemented with retirement homes, car parking facilities and so on in different localities.

However, such diversification is not without risk. In 1991 a number of construction companies profits were dramatically affected by their investment in property. The slump in commercial and domestic property values and in sales in the USA and Europe, resulted in companies reporting losses. Whilst diversification can spread the risk, the severity of the risk takes on a new importance.

Moreover, a company's systematic risk can be split into two components: business and financial risk. Business risk is the result of a company trading with its assets, financial risk arises directly out of the gearing process. Whilst the former is borne by the equity and debt holders, the latter brings risk only to the equity holders. The tradition that debt capital has priority over equity capital in both the distribution of the company's annual net cash flow (as interest and dividends) and in any final liquidation distribution, ensures that the risk of a reduced or zero dividend is borne only by shareholders. The severity of the financial risk is directly related to the level of gearing.

Whilst these are well established concepts they do not help us with the types of risk in the construction industry, which have been classified as follows:

Pure risk (sometimes called static risk)
> No potential gain: such risk will typically arise from the possibility of accident or technical failure.

Speculative risk
> The possibility of loss or gain which might be financial, technical, or physical.

Table 3 A shows some of the various types of risk relating to the construction industry. The items relating to the impact are discussed later in the chapter.

Table 3 A   The various types of risk relating to the construction industry

Event	Type of Risk	
Onerous conditions of contract	Speculative risk	Company risk
Exceptionally inclement weather occurring during the project	Pure risk	Company risk Project risk
Inflation causing a dramatic increase in the price of land	Speculative risk	Market/Industry risk
A national strike of all building workers	Pure risk	Market/Industry risk
Failure to find tenants for a speculative development at an economic rent	Speculative risk	Company risk
Failure by a building surveyor to identify a structural defect	Speculative risk	Company risk Individual risk
Injury caused by an accident at the work place	Pure risk	Individual risk

IMPACT OF RISK

The risk hierarchy

Figure 3.4 gives a simplified view of the risk hierarchy. At its broadest level, risk will have an impact upon the environment.

Figure 3.4   The risk hierarchy

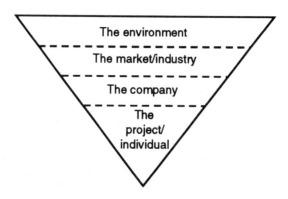

Risk and the general environment

The general environment is that which affects all organisations in a given society. General environmental risk can be divided into two parts. Firstly the physical and secondly the political, social and economic.

The physical environment includes the weather and other natural phenomena such as landslips and earthquakes. It can have a significant impact on the construction process. For instance, continuous rain may mean a change to the indoor and outdoor programme of work; high winds may cause changes to the structural steelwork erection programme; low temperatures may change the sequence of concreting operations. Whilst the physical environment cannot be controlled, the risks that arise from it can be identified and steps taken to mitigate their effects. For instance, re-scheduling particularly vulnerable operations for times when there is the greatest probability of favourable weather conditions.

The political, social and economic environment is partially controllable. The government can control events in the UK economy, but not in the world economy. Hence, in the international oil market, the British Government cannot dictate the oil barrel price for North Sea oil on the spot market. What characterises the political, social and economic environment is the speed of change, today more rapidly than ever before. For instance, within the space of ten years, Britain has moved from being a declining economic force to being one of the strong economies in the world. The wind of change has gained speed and it affects all industries.

The economic and social environment is controlled by government, but the various industries can be seriously affected by environmental decisions. An example would be the imposition of development controls on office buildings in city centres, which would have a severe impact on the workload of the development and construction industries.

A good example of the volatility of the environmental risk is the UK property crash in 1974. During 1973, in an effort to control the economy and prevent a balance of payments crisis, the Bank of England increased the minimum lending rate from 4.5% to 13%. Inflation rose rapidly and in early 1974, oil prices rose by 300%. The government introduced two pieces of legislation which led to the property collapse. A rent freeze was introduced and Development Gains Tax legislation imposed a tax on property development. Property prices collapsed and there was a loss of confidence in property companies.

The property crash resulted from a property boom followed by falling demand, a rent freeze, taxation, and a harsh monetary policy. The cause of the environmental risk was controllable, but only by government, and the effect on the property and construction industry was devastating.

Nobody can ignore environmental risks. Whilst pressure can be brought to bear upon government to influence or change decisions, in general, most events are uncontrollable by the individual or the company. In these circumstances, attention must be given to evaluating the risk exposure.

The market/industry risk

Market risk relates to any event that might affect the complete industry, such as a national strike of all building workers.

Throughout the developed world, the construction industry is characterised by a particular oligopolistic form. Typically, the industry contains a small number of relatively large firms and a very large number of small firms. The likelihood is that large firms will analyse their risk exposure in a more systematic way than the smaller firms.

All companies will want to ensure they maintain their market share of available projects, be they members of the design or the construction team. This means that they must constantly evaluate competition and price levels. All are subject to market risk, but since they are interdependent (through the oligopolistic structure of the market) the reactions of any one firm to market risk may have to take account of the likely actions of other firms in the industry.

The company risk

Any company operates within a market. The company will have a number of current projects at any one time, each project generally being a profit centre. Company risk and project risk are intrinsically linked because the company must ultimately bear the consequences of a risky project. Risk strategy, is therefore usually determined by group consensus.

To avoid over exposure to many risky projects, some companies will form a separate company for a particular project. For example, the construction of the Channel Tunnel is being undertaken by Transmarche Link which is a specially formed consortium.

Project risk and individual risk

Project risk and company risk are intrinsically linked. If a contractor has a major project which is losing money, it will have an effect upon the company's financial performance.

Many risks are easiest to see at the working level at or near the bottom of an organisation's hierarchy. At this level a project may consist of hundreds of items and activities. People operating at this level have a practical day to day understanding of the difficulties. It is impractical for people operating here to have a well developed overview of the total project. They are concerned about their project, they are rarely able to set their uncertainty into context by seeing the knock on effects with work packages on other projects.

Any risk management system has to recognise this by ensuring that it can be used equally bottom up and top down.

CONSEQUENCE OF RISK

When considering the consequences of a risk occurring, the relevant factors relating to effects of the risk are taken into account.

Most professionals will tend to rely upon expert judgement and knowledge, tempered with some information, if it is available, about past

> *Consequence*
> ☐ **the maximum probable loss**
> ☐ **the most likely cost of the loss**
> ☐ **the likely cost of servicing the loss if no insurance has been effected**
> ☐ **the cost of insuring against the event occurring**
> ☐ **the reliability of the prediction about the event.**

events. For instance, there will be statistics available for thefts of materials from sites in any year. But the contractor must balance the cost of the insurance premium with the excess he is prepared to carry on the insurance policy and the additional cost of security measures.

There are many sources of risks where no reliable data are available. Rather than ignore these sources, the course of action should be considered within the overall risk management system.

Table 3 B   Event - likelihood of damage to adjoining buildings as a result of pile driving

Severity \ Likelihood	Improbable	Rare	Possible	Probable	Very likely
Negligible (up to £100)	Retain	Retain	Retain	Retain	Retain
Small (£100 - £1,000)	Retain	Retain	Partial Insurance	Partial Insurance	Partial Insurance
Moderate (£1,000 - £5,000)	Retain	Partial Insurance	Insure	Insure	Insure
Large (£5,000 - £50,000)	Insure	Insure	Insure	Insure	Insure
Disastrous (over £50,000)	Insure	Insure	Cease activity	Cease activity	Cease activity

Table 3 B is an approach, albeit simplistic, to considering an event: the likelihood of damage being caused to adjoining buildings as result of piling on site. The table helps to consider the risk consequences in a structured way.

RISK ANALYSIS

The main purpose of a risk management system is to assist business to take the right risks. An integral part of the system is risk analysis. The wider and more efficient use of computers is likely to encourage more rigorous analysis of risks. The time has now arrived when the more significant risks can be evaluated with economic advantages.

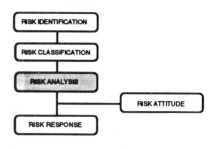

> 'To try to eliminate risk in business enterprise is futile. Risk is inherent in the commitment of present resources to future expectations. Indeed, economic progress can be defined as the ability to take greater risk.'
>
> *Hertz and Thomas (1984)*

Future chapters will discuss the tools and techniques of risk analysis. For instance, decision trees can be used to understand and determine the problem structures as part of the analysis.

The essence of risk analysis is that it attempts to capture all feasible options and to analyse the various outcomes of any decision. For building projects, clients are mainly interested in the most likely price, but projects do have cost over-runs and, too frequently, the 'what if' question is not asked.

Let's consider an example of an £800,000 extension and alteration to a school building. There are eight work packages, each costing about £100,000. However, the project is still not fully designed and they could vary by + or - 10% until firm bids are received. The project could therefore be between £720,000 and £880,000 which could be the best and worst cases. In reality, as long as the prices of the eight items are not linked together, the total is unlikely to fall outside the range of £760,000 to £840,000. The risk is not as bad as it first appears because it is unlikely that all eight packages will go wrong. Each of the packages needs to be analysed to identify the likely risks, these risks then have to be managed.

The use of risk analysis gives an insight into what happens if the project does not proceed according to plan. No matter how good the analytical techniques, it is up to the professional to interpret the results. When active minds are applied to the best available data in a structured and systematic way, there will be a clearer vision of the risks than would have been achieved by intuition alone.

STEP 1	All the various options should be considered
STEP 2	Consider the risk attitude of the decision-maker
STEP 3	Consider what risks have been identified, which are controllable and what the impact is likely to be
STEP 4	Measurement, both quantitative and qualitative
STEP 5	Interpretation of the results of the analysis and development of a strategy to deal with the risk
STEP 6	Decide what risks to retain and what risks to allocate to other parties.

Figure 3.5   Risk analysis

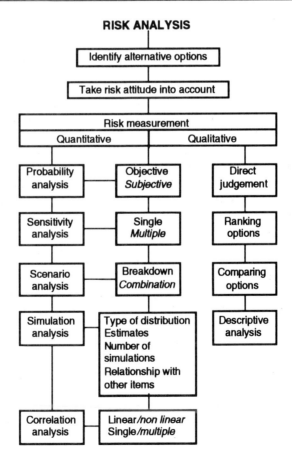

Figure 3.5 shows the sequence in risk analysis. The traditional approach to forecasting a construction price or construction duration at the design stage of a project, is to use the available data and produce a single point best estimate. The risk analysis approach explicitly recognises the uncertainty that surrounds the best estimate by generating a probability distribution based upon expert judgement. Therefore, the understanding about the effects of uncertainty upon the project will be improved. A more detailed discussion on risk analysis is in chapters 7 and 8.

An example of risk analysis

The effect of varying weather conditions is considered upon the excavation of isolated column bases in clay

Weather Conditions	Probability	Unit Price rate(£)	Probable cost	Time in minutes	Probable time
Very dry	.10	2.60	0.26	12	1.20
Fairly dry	.20	3.00	0.60	15	3.00
Wet	.50	6.00	3.00	25	12.50
Very wet	.20	8.00	1.60	35	7.00
		Probable cost	**5.46**	Probable time	**23.70**

The probable cost is £5.46 with a probable time of 23.7 minutes. The difference between the very wet and very dry unit price rates is considerable. The contractor must ask what premium should be added to the base unit price rate at which he still remains competitive in the market place yet gives some comfort that if the worst eventuality happens the loss suffered can be sustained.

The example assumes the contractor has no knowledge of the ground conditions, which is often not the case.

The analysis has not solved the problem, but it has highlighted the options. A discerning interpretation of the results is required.

RISK RESPONSE

The response to, or the allocation of, risk can take any of four basic forms, as shown in **figure 3.6**.

Figure 3.7 shows the approach to risk response on a construction project.

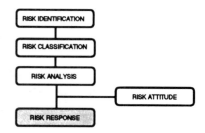

Figure 3.6   Risk response

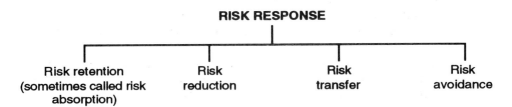

Proper allocation of risk must consider the ability to absorb the risk
and the incentives being offered to carry it. For instance, when pricing for
an underground construction project, the contractor will have available the
site investigation and bore hole reports as well as any geological
information. The specification calls for the contractor to make due
allowance for excavation in any ground condition he encounters. Incomplete
knowledge of the site geology and the possibility of unforeseen ground
conditions poses great risks to the contractor. Part of that risk will have to
be transferred to the works contractors. However, a defensive position will
be taken by adding a risk premium and inflating the unit price rates, thus
hopefully covering the risk being retained. In this situation there is some
transferred risk and some retained.

Figure 3.7   Risk on construction projects

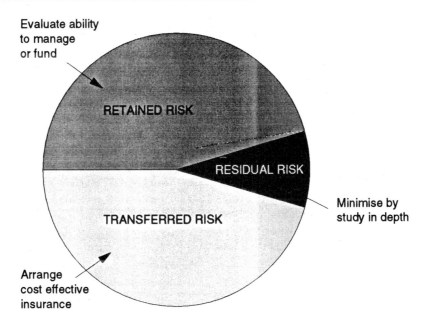

On lump sum contracts, clients are passing more risk to contractors and trade contractors. We have to be explicit and courageous when we take risks and not just gripe about them. Clients have to accept that risk and reward go hand in hand. Risk response actions must not be viewed as wasted costs, but as investments that generate returns. Listed below are some of the fundamental principles which govern the allocation of risk.

Some fundamental considerations which govern the allocation of risk:
- ❏ **which party can best control the events that may lead to the risk occurring;**
- ❏ **which party can best manage the risk if it occurs;**
- ❏ **whether or not it is preferable for the client to retain an involvement in the management of the risk;**
- ❏ **which party should carry the risk if it cannot be controlled;**
- ❏ **whether the premium to be charged by the transferee is likely to be reasonable and acceptable;**
- ❏ **whether the transferee is likely to be able to sustain the consequences if the risk occurs;**
- ❏ **whether, if the risk is transferred, it leads to the possibility of risks of a different nature being transferred back to the client.**

Risk retention

Risks that produce individually small, repetitive losses are those most suited to retention. On a motor insurance policy many people will bear the cost of the first £50 in return for a reduction in the premium. Some people may choose to pay the first £200 in return for a larger reduction; the level of retention is dictated by the financial circumstances and the likelihood of loss.

Not all risk can be transferred, but even if they were capable of being transferred it may not prove to be economical to do so. The risk will then have to be retained. Besides, it is preferable to retain a portion of the risk in certain circumstances. For example, a reduction in an insurance premium with a corresponding retention in the form of limited excess provision in the event of a claim, may be preferred to full coverage; the gamble is between paying the premium and the probability of the event occurring and the consequential loss that would result. Many people take that risk every day by insuring their cars for third party, fire and theft in preference to fully comprehensive insurance cover. It might be they cannot afford the premium, or the value of the car does not warrant the cost of the premium, or they may feel they are such a good driver that there is a very low probability of them causing an accident and being held responsible.

When considering professional indemnity insurance, the professional will evaluate what excess they feel it is appropriate to retain. There are some design practices in London who carry the first £500,000 of each and every claim made against them for negligence. Their risk retention is based on the probability of a claim for damages and what the practice can afford to pay if the claim is successful.

In summary the relevant factors are:

- ❏ the cost of the insurance premium;
- ❏ the maximum probable loss;
- ❏ the likely cost of the loss;
- ❏ the likely cost of paying for the loss, if uninsured.

Risk reduction

One of the ways of reducing the risk exposure is to share risks with other parties. For instance, the international banks' syndicated loans to Third World governments or the stock market investor spreads his risk over many securities. The general contractor will attempt to reduce his exposure to pay liquidated damages for late completion by imposing liquidated damages clauses in domestic sub-contract agreements.

Similarly with the contractual arrangement, the use of management fee types of contract will remove the adverse attitude of contractors and should reduce the likelihood of claims from the contractor for direct loss and expense.

Risk reduction falls into four basic categories. Firstly, education and training to alert the staff to potential risks. Secondly, physical protection to reduce the likelihood of loss. For example, a contractor could employ an independent quality assurance company to act as a second check on all its projects - it's an expensive option but it would reduce the incidence of defects going undetected. Thirdly, systems are needed to ensure consistency and to make people ask the 'what if' questions. Finally, physical protection can be taken to protect people and property.

In a building the typical case of risk reduction would involve the installation of a sprinkler system. Whilst the regulations might not require the sprinklers, the client would feel the need to reduce his likelihood of loss from fire damage by paying for the sprinkler installation.

Risk transfer

Transferring risk does not reduce the criticality of the source of risk, it just removes it to another party. In some cases, transfer can significantly increase risk because the party to whom it is being transferred, may not be aware of the risk they are being asked to absorb. For example, the general contractor, when entering into a contract with a sub-contractor, may impose a liquidated and ascertained damages clause for late completion which includes both the damages for the main contract and an assessment of loss suffered by the general contractor. The sub-contractor may not be aware of the additional risk being transferred and he may well not be in the financial position to carry such risk.

The commonest form of risk transfer is by means of insurance which changes an uncertain exposure to a certain cost. In the construction industry obtaining insurance cover is becoming more expensive. On construction projects, fault-free building cannot be guaranteed and defects may be discovered long after practical completion. Latent defects which cannot reasonably be discovered at the stage of a building's practical completion or during the contractual defects liability period are a fact of life.

Current arrangements for handling the unpleasant consequences of discovery of latent defects fail to satisfactorily serve the interests of either clients, contractors or designers. The responsibilities and liabilities of all involved in the building process are often needed to be established in court. There is a risk that clients would have to prove, through legal processes, that the defect and resulting damage were caused either by breach of contract, negligence or omission under tort in order to recover the costs of litigation and repairs. Clients might not even have sufficient funds to initiate the legal action. As for contractors and designers, they are potentially liable for claims many years after practical completion. Moreover, the application of joint liability in multi-party actions may lead to one or more members of the construction team having to accept an unfair share of the compensation to be paid.

Furthermore, professionals are exposed nowadays to the floodgates of liability in an indeterminate amount for an indeterminate class of people. In a way, the availability of professional indemnity insurance has encouraged the courts to increase the exposure beyond the reasonable limits of the notions of 'reasonable care and skill' and the 'neighbourhood relationship'. Moreover, legal procedures are often multi-party and long drawn out, demanding the time of senior staff.

Although professional indemnity insurance has been a loss maker for insurers, the aggregate cost of administrative fees and litigation are very high compared to that devoted to building repairs - the real but indirect purpose of professional indemnity insurance.

It has been suggested that insurance may increase the risk of a claim being made against the insured. If something goes wrong on a project, the entrepreneur may look for some way to recover the losses. There is the view that the legal liability seems to be falling on the person most able to pay.

Various reports have recommended that a project insurance, based on first party material damage without proof of fault, would be the most advantageous for the client. The project insurance policy should be negotiated by the developer or building owner at the preliminary design stage. During its currency, the policy would be assignable to successive owners and whole-building tenants. The major advantage will be the safeguard for clients and tenants against the costs and consequent worries arising from discovery of latent defects covered by the recommended policy. Cost of repairs will certainly be recoverable as it facilitates making good insured defects and damage both speedily and efficiently on an equitable basis. The responsibilities and liabilities of all involved in the building process are also made clear without recourse to litigation. Dependable building quality will then be promoted without stifling innovations through fear of litigation or threat of excessive insurance costs.

Withholding retention money on interim payments to the contractor and specialist contractors is a way of covering residual risks that may arise. The sums are held to ensure the contractors complete their work properly and to cover the risk of the loss arising from the liquidation of the contractor. The amount of retention withheld does vary, but is usually 3% - 5% of the project value. An alternative is bonding, particularly used in the USA. A performance bond provided by an insurance company or a bank, ensures the project will be completed in the event of default by the contractor. Bonds are a form of insurance which is purchased and ultimately paid for by the client.

Risk retention and risk avoidance, shared confidence and shared fears

---

Risk avoidance

Risk avoidance is synonymous with refusal to accept risks. The refusal to contract is a simple example of risk avoidance. However, it is more relevant to consider the specific risks which can be avoided. Normally, risk avoidance is associated with pre-contract negotiations but it may well be extended to decisions made in the course of execution of the project. For example, when a contractor has committed a fundamental breach, the employer is entitled to rescind the contract. However, the employer can waive this right of rescission and thus allow all the concomitant risks under the contract to continue. If the employer rescinds, he is in fact avoiding any further risks by accepting that the contractor has repudiated the contract. A more pertinent example of risk avoidance is the use of exemption clauses, either to avoid certain risks or to avoid certain consequences flowing from the risks.

RISK ATTITUDE

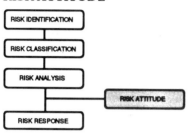

Most people think of great risk as an unpleasant thing, something to be avoided if possible. There is much evidence of people who are more reluctant to undertake uncertain projects than certain ones. This attitude may not be shared by all people, but it is easy to rationalise.In simple terms, there are three types of people/organisations:

- ❑ risk loving
- ❑ risk averse
- ❑ risk neutral.

The classification into two groups is a gross oversimplification. The image of the corporate executive as a bold risk-taking wheeler-dealer is part of the folklore of business. Rupert Murdoch's News International Corporation borrowed US$8.3 billion to launch Sky Television and to buy newspaper companies around the world. The company was highly geared but could only afford to pay the interest on loans as long as the economy was buoyant. The world recession came and business slumped resulting in the debt being re-scheduled. Rupert Murdoch could be described as risk loving, but that policy has paid off handsomely in the past. The evidence is that, despite the turbulence, Sky Television will be highly profitable following the merger with British Satellite Broadcasting.

The risk lovers are still in business, but they are harder to find. Most decisions are now made on the basis of detailed analysis. Even detailed analysis however cannot prevent a bad decision being made. Rolls Royce's decision to accept the ill fated RB211 jet engine contract with Lockheed was an instance where managers were unsure of how to incorporate the project risk into the company's strategic risk. Rolls Royce subsequently went into liquidation only to be saved by the British government.

The basic rationale of risk aversion is that it is more unpleasant to lose a given sum than pleasant to gain the same amount; for a loss may cause a reduction in the standard of living to which one has become accustomed. In an extreme case, a risky venture may be shunned because it puts basic necessities at risk.

SUMMARISING RISK MANAGEMENT

A risk neutral person is one who treats risk and reward on an equal basis. The concept of utility and risk attitude is discussed in more detail later. A summary of the risk management system is as follows:

- ❏ risks have to be identified, classified and analysed before any response is made;
- ❏ an identified risk is not a risk, it is a management problem;
- ❏ beware of using solely the intuitive approach or 'gut feel' to manage risk;
- ❏ risk management needs to be continuous from the moment the project starts to the moment it ends;
- ❏ a poorly defined risk structure will breed more risk;
- ❏ use both a wide angled lens and a zoom lens for your vision of what could happen in the future;
- ❏ use both creative and negative brainstorming, don't use the ostrich approach;
- ❏ always have a contingency plan to cope with the worst eventuality;
- ❏ risk management systems should not be complicated or burdensome, they need to be integrated into a firm's daily operations.

THE LADY OR THE TIGER

Three young men could open either door they pleased. If they opened the one, there came out of it a hungry tiger, the fiercest and most cruel that could be procured, which would immediately tear them to pieces. But, if they opened the other door, there came forth a lady; the most suitable to her years and station that His Majesty could select from among his fair subjects. So I leave it to you, which door to open?

The first man refused to take the chance. He lived safe and died chaste.

The second man hired risk management consultants. He collected all available data on lady and tiger populations. He brought in sophisticated technology to listen for growling and detect the faintest whiff of perfume. He completed check lists. He developed a utility function and assessed his risk attitude. Finally, sensing that in a few more years he would be in no condition to enjoy the lady anyway, he opened the optimal door. And was eaten by a low probability tiger.

The third man took a course in tiger training. He opened a door at random and was eaten by the lady.

(Taken from W C Clark - Witches, Floods and Wonder Drugs: Historical Perspectives on Risk Management.)

4

SOME OF THE TOOLS AND TECHNIQUES OF RISK MANAGEMENT

INTRODUCTION

This chapter discusses some of the tools and techniques for analysing risk and making decisions under risk. There are many ways of doing this, from the fairly simple to those which require a computer as a minimum tool. We shall consider techniques with varying degrees of complexity, concentrating upon the principles of the various methods rather than their detailed application.

Risk management is one aspect of management science. There are two broad categories of management science techniques: *deterministic*, and *probabilistic* or *stochastic*. Deterministic techniques assume that the values of the decision variables are known with 100% certainty, which is rarely the case with construction.

Probabilistic or stochastic techniques on the other hand, are concerned with factors that cannot be estimated with certainty, such as most data associated with construction.

For the most part decision-makers tend to concentrate on single values of outcomes, such as profit, which have been calculated from single value estimates of the variables. This approach does not distinguish between high and low risk projects. A decision-maker may or may not be in a position to control the uncertainties, but he should be in a position to make a quantitative estimate of the risk involved in any estimate.

Whilst most of the tools and techniques discussed in this chapter provide quantitative solutions, they will incorporate some subjectivity. Chapter 3 discussed the importance of judgement and experience. However, the analyst should rely more heavily on the output of a quantitative forecast rather than on intuitive judgement.

An important source of bad decisions is the *illusion of certainty.*

There is no doubt that judgement stimulates thought and explores new relationships, but, where possible, quantitative techniques should be incorporated to test and support assumptions. Most importantly, always recognise what the statistical technique is doing with the data.

Statistics are like a bikini suit - what is revealed is interesting, what is concealed is crucial.

(*Feinstein*)

The tools and techniques of risk management help us to determine a decision. Determining has two senses; finding out and assuring. Once a situation has been determined it requires action. The tools cannot make the decision, only humans can initiate the course of action.

Seeing the big picture and the detail

When we make decisions we need to have clear objectives, goals, plans and strategies. We need two types of camera in our mind; one with a broad angle lens that can see the breadth of the whole problem, but not in great detail; the other with a zoom lens that can zoom in on specific risk areas which require more in-depth analysis.

Decision-making techniques	*Where they are used*
❑ **The risk premium**	**Risk response**
❑ **Risk-adjusted discount rate**	**Risk response**
❑ **Subjective probability**	**Risk response**
❑ **Decision analysis**	**Risk analysis/**
Algorithms	**Risk classification**
Means-end analysis	
Decision matrix	
Bayesian theory	
❑ **Sensitivity analysis**	**Risk analysis**
❑ **Monte Carlo simulation**	**Risk analysis**
❑ **Portfolio theory**	
❑ **Stochastic dominance**	

Decision-making techniques

A number of decision-making techniques can be used in risk management and investment appraisal, as shown in the list above.

These are now considered in turn.

THE RISK PREMIUM

A discount rate reflects the investor's time value of money <u>and</u> the rate of return the property must earn to justify the investment. Some projects are almost risk free, such as the case of a tenant prepared to sign a long term lease on a building at the early design stage, whereas a project which is considered risky will attract a premium to the discount rate.

The investor in land and property will balance the costs and the revenue of the investment over a period of time by using a discount rate. Similarly the contractor and specialist contractor are looking at the investment of their resources and effort into a construction project: there is a risk of loss which is tempered by the possibility of gain. They might use discounted cash flow techniques (or a non-discounted approach) in order to evaluate the project.

The risk premium will be added to the risk free discount rate. The risk premium will vary with each project, depending upon attendant risk and the attitude towards risk taking. For example, building an office block in the Mayfair district of London is likely to be a safer investment than building the same block in the docklands. A discount rate to be used for the future cash flows for the Mayfair project might be 11% whereas for the docklands project the rate might be 15%. The additional 4% is a reflection of the risk premium the investor feels is appropriate.

There are no formulae which derive an appropriate risk premium; each investor will have his or her own requirements as to the risk premium for each project.

The difficulty in choosing the appropriate risk free discount rate is rooted in the inaccuracy of the term *risk free*. Financial commitments always carry certain risks which can be neither eliminated nor transferred. The term risk free is intended to imply not absolute absence of all risk, but virtual absence of default risk. In financial terms, the risk free rate is taken as that which would apply if lenders viewed a borrower's credit and collateral so favourably that they were absolutely certain of repayment at the scheduled time.

A contractor or a specialist contractor will also add a risk premium to a project, although they might not identify it as such. When pricing an estimate for a tender, if there is information or a party to the contract they are unsure about, they will add an additional margin for unforeseen eventualities.

The determination of the appropriate risk premium is based upon the trade off between risk and return. Unfortunately the concept of discounting embodies the precept that the further into the future the risk premium is applied, the greater is the impact on the present value of an investment. In other words, future risk is discounted more heavily than is near-term risk. A single risk-adjusted discount rate, therefore, is a poor proxy for the impact of risk on value over the project's lifetime, because risk does not increase exponentially with the passage of time. Arguably, the greatest uncertainty surrounds the initial construction period. Once the project has been built and tenants have been found, the risk for a client starts to diminish. Obviously, running costs are still a concern but after a period of occupancy they are easier to forecast. One way of solving this difficulty is to make appropriate adjustments to the risk premium for the varying risks at different stages of the project. This approach is similar to varying the allowance for inflation in the discount rate for differing time periods.

The issue of risk is thus very important for profitability in the industry. But how do decision-makers decide on the risk premiums and the riskiness of earnings and costs? That depends upon their attitudes to risk and upon the probabilities associated with an event. What we are concerned with here is the framework within which these variables can be handled.

RISK-ADJUSTED DISCOUNT RATE

It is tempting to consider the risk premium as the requirement for an additional rate of return. In other words, it has been suggested that one way of taking risk into account in investment appraisal is to use risk-adjusted discount rates. A real discount rate used in say, life-cycle costing calculations, may be viewed as composed of three parts: a time value of money; an adjustment for expected inflation; and a risk premium.

The risk premium is added to reflect the investor's view of the degree of sensitivity of the project to risky factors. The size of the premium depends upon the degree of risk associated with the project and the attitude to risk by the investor. The greater the risk, the greater the premium. In practice, a single risk-adjusted discount rate is added to the discount factor:

$$RA = (RF + I + RP)^t$$

RA = Risk-adjusted discount rate
RF = Risk free rate
I = Allowance for inflation
RP = Risk premium which is the adjustment for extra risk above the normal risk

A potential disadvantage of this approach has already been noted. Since the discount factor is part of a compounding function, the discount factor grows with increases in the value of t. This implies a special assumption that the risks associated with future costs and revenues increase geometrically with time. This assumption is usually justified on the grounds that the accuracy of our forecasting decreases with time.

A procedure for evaluating such projects is to separate timing and risk adjustments using the concept of certainty equivalent value (CEV). The CEV of a cash flow in a given year is simply its risk adjusted value in that year. Hence, if all future cash flows were converted to CEVs, they could then be discounted to the present using a single risk free discount rate. With the timing and risk adjustments separated, the risk adjustments can then be viewed more analytically.

If a company uses the same discount rate for all projects regardless of their different risks, it will tend to reject low risk projects that are profitable and invest in some high risk projects that are unprofitable. For this reason, many companies use different discount rates for different kinds of assets depending upon their risks.

An alternative is to discount cost and benefit streams separately, each with a unique risk-adjusted discount rate (RA). As future **benefits** become more uncertain into the future, the RA is adjusted upwards as perceived risk increases. However, for **cost** streams, the RA is adjusted downwards as perceived risk increases, which means as future costs become more uncertain, so this is reflected with the lower rate.

A practical example of this might be in the construction of a bowling alley. Bowling has been undergoing a renewal of interest over the past five years. Over a 10 year time horizon it is difficult to know how the revenue streams will be maintained in the later years; hence the later years have a higher risk associated with them. Similarly, the cost streams may be affected by the need for additional refurbishment to cater for changes in taste over the later years, adding to the risk in the later years of the time horizon.

Most companies using risk-adjusted discount rates employ a risk classification scheme. Typical projects are classified into, say, four categories. Each has a different level of risk and a different risk premium. If there are only four risk classes, there are only four discount rates used at a time.

Figure 4.1 shows a graphical representation of a typical relationship between required rates of return and risk classes.

The reason for using risk classes is that it minimises bias in project evaluation. We need to identify why we are assigning a much higher premium to a build/operate/transfer project compared with a housing scheme.

Figure 4.1   Rates of return and risk classes

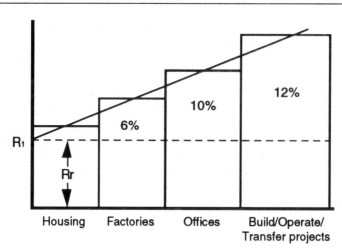

SUBJECTIVE PROBABILITIES

Probability analysis has been deemed to be useful when decision-making is characterised by risk rather than uncertainty; that is, where there is a history of outcomes that can be aggregated to produce objective probabilities. However, many business decisions by their nature may not be able to satisfy the kind of information requirements described above. Given the definitions in chapter 2, these decisions are uncertain. However, uncertain situations can be changed to risky situations simply by the assignment of subjective probabilities. This is not so much of an arbitrary procedure as one might think at first sight. There are a number of good reasons why subjective probabilities may be not only quite acceptable but even preferable to objective probabilities. Koutsoyiannis (1982) makes the point that most of the decisions which a firm must take are unique, in the sense that the conditions of the economic environment change continuously. It is impossible to obtain past observations of similar events from which to estimate objective probabilities.

Objective probabilities then, may be inappropriate for projects simply because of the information requirements and the fact that like must be compared with like. It might be felt to be especially true that objective probabilities are difficult to obtain in the construction industry, where many buildings are one-off projects. Nevertheless, an extensive historical record exists, and a wealth of experience has been developed. Past experience is very useful for investment appraisal. In particular, past experience and learning is invaluable in the definition of the subjective probabilities that characterise specific decisions.

Subjective probabilities may be important for a second major reason. The subjective probabilities of different individuals with the same experience and information may be very different. Personal differences of the decision-maker can and do play an important role in decision-making. Subjective probabilities explicitly recognise this at an early stage in the analysis. There are some problems of course; a subjective probability is a measure of the degree of confidence of a particular individual in the truth of a particular position. This implies that if we simply asked for the probabilities from a manager, investor or entrepreneur, we might be in danger of prejudging the outcome. Subjective probability, in other words, might generate a self-fulfilling prophecy.

Given that subjective probability is a measure of the degree of confidence of an individual in a position, it is a rather arbitrary measure of risk. It is likely that the engineer or contractor who is at the centre of the proposals may have a vested interest in the project. Consciously or not, the individual concerned may well bias the analysis. This is a problem for many firms engaged in many different industries. In construction, with a high degree of uncertainty and variability in performance of buildings of a similar type, there is a large scope for such bias.

The decision to invest will, in the end, be a subjective decision based on the experience of the board and so on, but there are ways to do this effectively such as the *Delphi Method*. The role of risk analysis is to be a reasonably objective guideline to aid decision-making rather than merely to provide high level management with a measure of the degree of confidence of some individual located somewhere in the organisation who may favour the project.

Subjective probabilities are not pulled out of thin air. They will usually be based upon knowledge and experience gained from similar projects. Even the most extreme one-off has some connections with past projects and current technologies. In addition, subjective probabilities should not be formed by one individual or group. They should be subjected to testing and questioning through tried and tested techniques such as the *Delphi Method*. If this is done then subjective probabilities will acquire many of the desirable properties of objective probabilities. In other words, so long as the dangers are recognised, they can be avoided.

DECISION ANALYSIS

Decision analysis is a technique for making decisions in an uncertain environment that formally treats both risk exposure and risk attitude. It provides a methodology to allow a decision-maker to include alternative outcomes, risk attitude and subjective impressions.

Decision analysis deals with the process of making decisions. It is both an approach to decision-making and a set of techniques to guide decision taking under conditions of risk and uncertainty. These decisions may be long-term, strategic, or short-term decisions affecting one particular area. They may be opportunities to exploit a chance to enter a new property market or plan a new development. They may deal with a anticipated difficulty or they may be crisis decisions.

Decision analysis follows a number of steps:

- ❑ recognising and structuring the problem;
- ❑ assessment of the values and uncertainties of the possible outcomes;
- ❑ determining the optimal choice;
- ❑ implementation of the decision.

The decision techniques considered in this section are fairly simple. They are:

- ❑ algorithms;
- ❑ means-end chain;
- ❑ decision matrix;
- ❑ decision trees;
- ❑ stochastic decision tree analysis.

Algorithms

An algorithm contains a sequence of instructions for problem solving. These are the steps in a task and the responses determine the route to be followed. Algorithms have often been used as a prelude to computer programs as they are logical and easy to follow. The example below is self-explanatory.

Figure 4.2   A simple algorithm used for fault diagnosis

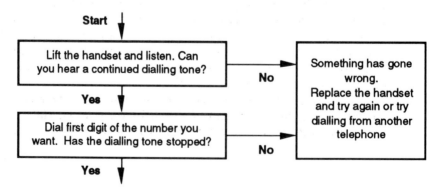

Figure 4.3   A means-end chain

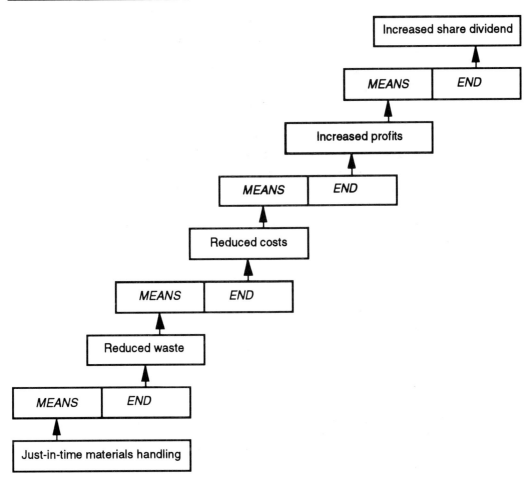

Means-end chain

This is a method of clarifying a chain of objectives and therefore identifying a series of decision points. It relies on the fact that what is an objective to one decision-maker might be a means of achieving a higher objective to a higher level decision-maker. The analysis is carried out by charting a means-ends chain as shown in **figure 4.3**. The example is that of an improvement in materials handling.

The means-end chain link is shown in **figure 4.4**. The hierarchical nature is suggested by presenting the chain in a step-wise fashion to indicate that the means operates at a lower level to achieve a higher end.

Figure 4.4   A means-end chain link

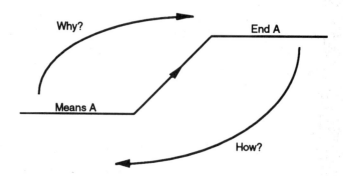

The simplest means-end relationship need be carried no further if it is possible to directly employ means A in order to achieve end A. However, most problems are multi-level and require a multi-level means-end chain as shown in **figure 4.5.**

Figure 4.5   A multi-level means-end chain

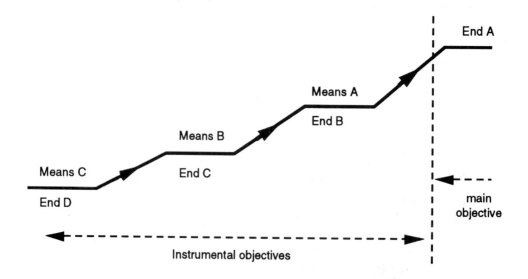

In the example, having established end A, the next step is to pose the question 'How can end B be achieved?' End B is an intermediate goal which may still be too difficult to directly accomplish, so the decision-maker seeks a means B by which to achieve this end. This process may continue for several levels until a point is reached at which an impossible means is required.

Figure 4.6 shows an example of a pre-fabricated timber frame housing manufacturer who recognises that the houses must be priced at US$350 per sq m. in order to achieve a substantial market share which will allow the manufacturing plant to operate at 95% capacity. Because of inefficiencies, the present plant is working below capacity and the quality of the product needs to be improved.

Figure 4.6   The means-end chain used by a timber frame housing manufacturer

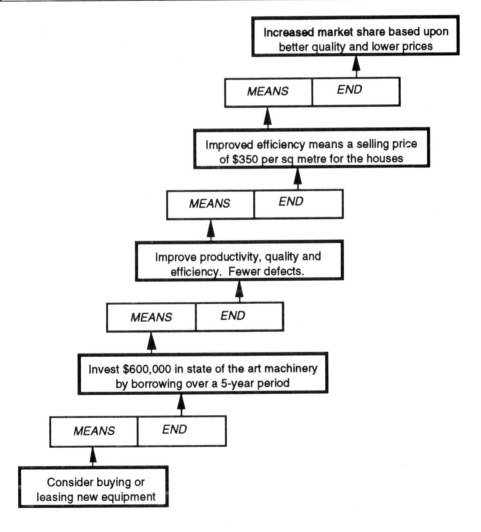

The end is a reduction in the selling price in the market place and an improvement in the quality generating an increase in the market share. The means is an investment of US$600,000 to improve the product and the quality.

The means-end chain described here is very simple, with a one-to-one relationship. In practice, the development of such a chain results in a variety of combinations in which a given end may have several means. The result is a highly complex set of relationships in which there may be competitive means and ends. For example, in **figure 4.6**, instead of investing US$600,000, another choice would have been to have sold the plant to a more successful or efficient competitor.

The means-end chain simply specifies the sequence, alongside this the risk should be considered by taking into account the most likely, the best and the worst eventualities.

The means-end chain is a formalisation of techniques we have always used, but it does encourage the orderly cataloguing of goals and their placement in a hierarchy. It also identifies gaps in the relationship between means and ends.

Decision matrix

A decision matrix is a representation of the options that are open to the decision-maker, the factors that are relevant, and the outcomes. This can be illustrated using a simple example. The Pulsometer Pump Company makes concrete pumping equipment for the European market and is at present operating at near production capacity. The sales and marketing director of the company anticipates that the market for concrete pumps will increase by 15% during the next twelve months. The board must decide how to react to this change in demand. There are three strategies that are being considered:

Strategy
S1 Install new equipment to improve productivity with a new system of working
S2 Institute overtime and weekend working
S3 Continue to work at capacity and let rivals or new firms satisfy the increased demand.

The contribution that each strategy will make to profits over the period of the next twelve months is estimated. Installing new equipment leads to a profit forecast of £240,000, overtime working leads to a profit of £210,000, and continuing to work at near capacity would yield £170,000 over the period. These values are based on the assumption that the market grows by 15%. However, the marketing director admits that two other outcomes are also possible: that demand falls by 10% as a result of a new high technology Japanese pump entering the European market; or that demand remains unchanged because of a lack of growth in construction demand.

A decision matrix can be constructed to show this information. The options or strategies are shown in the rows and the factors or states of the market are shown in columns in **figure 4.7**.

Figure 4.7   Decision matrix (£'000s)

Strategy	Market factors			
	15%	**stable**	**-10%**	
S 1	240	130	0	
S 2	210	150	70	**Profits**
S 3	170	150	70	

What decision should the board take? Probabilities can be assigned to the various outcomes from which expected values can be calculated. For the Pulsometer Pump Company we need the probabilities associated with the change in market demand.

Market Outcome	Probability
15% rise	0.6
Stable	0.3
10% fall	0.1

The expected values for a particular strategy are obtained by multiplying the pay-off for each market outcome by the probability of that outcome and summing the results. Note that the probabilities are a subjective judgement. Thus for strategy 1, installing new equipment, the expected value is shown below.

Strategy 1

Outcome	Pay-off	*	Probability		Expected monetary value
15% rise	240,000		0.6	=	144,000
Stable	130,000		0.3	=	39,000
10% fall	0		0.1	=	0
	Total expected monetary value (EMV)			=	183,000

The same process can be repeated for strategies S2 and S3. This yields the respective expected values for all three strategies.

Strategy		Expected monetary value (EMV)
S1	New equipment	183,000
S2	Overtime	178,000
S3	Existing level	154,000

This leads us to favour approach S1 because it has the highest expected monetary value. However, other factors should be taken into account in comparing these strategies.

The variability of returns is important and can be used to give a further measure of the degree of risk. This measure of risk can be found by using standard deviations. The variance in statistics is given by the sum of the squared deviation of each pay-off from its expected value multiplied by the probability of that outcome. Hence, for strategy S1 the standard deviation is:

Strategy 1

Outcome	EMV	Deviation (D)	D^2	Probability		Total
240	183	+57	3249	(0.6)	=	1949
130	183	-53	2809	(0.3)	=	842
0	183	-183	33489	(0.1)	=	3348.90
				Variance	=	6140.30
				Standard deviation	=	78.36

Now we have two pieces of information: the expected value and the standard deviation. Since higher expected values will tend to be associated with higher values for the standard deviation, we should use instead the coefficient of variation which shows the proportionate deviation.

$$\text{Coefficient of variation} = \frac{\text{Standard deviation}}{\text{Expected value}}$$

Strategy	Expected value	Standard deviation	Coeff. of variation
S1	183	71.37	0.39
S2	178	44.50	0.25
S3	154	32.34	0.21

S1 has the highest expected value but is the most risky strategy. The choice of strategy will then depend upon the trade-off the decision-maker wants between return and risk. A highly risk averse decision-maker, for example, may prefer the 'do nothing' strategy S3 of continuing to work at capacity and letting the competitors satisfy any increased demand. Note that this is an example in which much of the primary information is based upon subjective estimates: profit forecasts and the probabilities of the three market outcomes. The final decision is, therefore, in part subjective but, by using the decision matrix, is based upon objective criteria.

Decision trees

A decision tree is a means of setting out problems that are characterised by a series of either/or decisions. It shows a sequence of decisions and the expected outcomes under each possible set of circumstances.

Most decision problems, including those in the construction industry, have some definable structure. By adopting an analytical approach such as a decision tree, the decision-maker is forced to recognise the existence of certain basic elements in the structure. There will be a set of objectives whose attainment depends on the decision-maker. For example, the client may want the building operational at a particular date, in which case he may be prepared to pay more on the construction price to achieve the certainty of completion. The decision-maker has to keep an open mind to search for all the available options for any decision.

A measure of the value for each possible outcome is required in order to give meaning to the decision tree. The most frequently used is the expected monetary value (EMV), which is the sum of the payoffs (or values) weighted by their probabilities. The decision-maker has to place a probability on a particular outcome. The probabilities are likely to be rough assessments and hunches, but most importantly, the use of a decision tree gives a structured approach to setting out a decision strategy.

Figure 4.8 shows a simple example for a decision tree. The general contractor has three options when tendering for three projects. He has the resources to undertake only one of the projects and must select the most profitable option.

The first option is to act as a general contractor submitting a lump sum bid for the re-building of a sea wall. The project has a likely profit of £400,000, but there is a chance that it could show a loss of £200,000. The second project is a design and build scheme for a new pumping station at a waterworks. The potential profit is £220,000, but again the project could show a loss, this time £100,000. The third project is a management contract for the refurbishment of an aircraft hanger with a potential profit of £160,000 but the possibility of a loss of £20,000.

Reading the diagram from right to left, the contractor has put probabilities associated with the profit and loss for each project. The cost

Figure 4.8   A decision tree

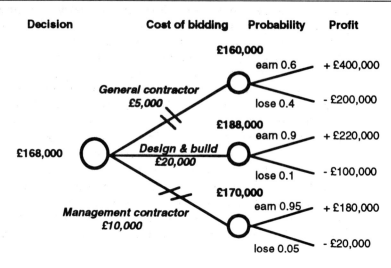

of bidding for the project is then deducted from the EMV to identify the project within the highest net EMV, which is the design and build scheme.

Consider, for example, the sea wall.
EMV = £400,000 x 0.6 - £200,000 x 0.4 = £160,000
less the bidding costs of £5,000 giving a net EMV of £155,000.

Decision trees are extensively used in managerial decision-making. They have the distinct advantage of forcing the decision-maker to structure the problem with which he is dealing in a consistent and objective manner. They then show which decisions are dominated by the other choices: in this case the option of building a sea wall is clearly dominated by the remaining options.

Bayesian theory

> **Thomas Bayes was born in 1702. He was a Presbyterian minister from 1731 to 1752 in Tonbridge Wells. He wrote 'An essay towards solving a problem in the Doctrine of Chances', which was not published until after his death in 1761. His fame was posthumous and is essentially a 20th century phenomenon.**
>
> *(Open University)*

Using the theory of probability developed by Bayes, we can examine the sensitivity of decisions. Bayes developed a theory of prior and posterior probabilities: prior probabilities can be revised in the light of additional information to form posterior probabilities. Hence, Bayesian theory is used to incorporate new information into the analysis.

For example, the Pulsometer Pump Company is approached by a firm of consultants who offer to survey the market for concrete pumps. The survey will cost the company £5,000. Should the board choose to pay for this additional information? Doing so will lead to a revision of their prior probability estimates of the likely market outcome.

The company enquires about the consultant's track record. The findings are tabulated below. It shows that in periods when the market did increase, 70% of the consultant's reports indicated a rise in the market. At the same time, when the market increased, 20% of reports indicated that the market would be stable and 10% indicated that the market would fall. The other figures in the table below have similar interpretations.

Actual Market Outcome	Prior Probabilities	Consultant's Prediction		
		Rise	Stable	Fall
Rise	0.6	0.7	0.2	0.1
Stable	0.3	0.2	0.6	0.2
Fall	0.1	0.1	0.2	0.7

Bayes theorem is used in order to incorporate this information to amend the prior probabilities. Assume that there are r mutually exclusive events E_i $(i = 1.....r)$ with prior probabilities $P(E_i)$. Assume further that there are events F^k and that the probability of F^k given that E_i has occurred is $P(F^k/E_i)$. Then the probability that E_i will occur if we know that F^k has occurred is given by:

$$P(E_j / F^k) = \frac{P(E_j) \times P(F^k / E_j)}{\Sigma\{P(E_i) \times P(F^k / E_i)\}}$$

If there are i mutually exclusive events E_i $(i = 1.....r)$, an event F can occur only if one of these r events happens and the probability of E_j happening when F is known to occur is:

$$P(E_j / F) = \frac{P(E_j) \times P(F / E_j)}{\Sigma\{P(E_i) \times P(F / E_i)\}}$$

$$P(E_i) \quad = \quad \text{Prior probability of event } E_i$$

$P(F^k/E_j) \;=\;$ Conditional probability of outcome F^k given E_j has occurred.

$P(E_i/F^k) \;=\;$ Posterior probability of event E_j given F^k has occurred.

In our example the prior probabilities of each outcome were:

$$
\begin{aligned}
E_1 - \text{rise} &\quad \rightarrow \quad P(E_1) \;=\; 0.6 \\
E_2 - \text{stable} &\quad \rightarrow \quad P(E_2) \;=\; 0.3 \\
E_3 - \text{fall} &\quad \rightarrow \quad P(E_3) \;=\; 0.1
\end{aligned}
$$

If F^r is a survey showing a rise in the market, then from the table we know that when the report predicted a rise in the market, the outcomes were: a rise at 0.7, stable at 0.2, and a fall at 0.1.
Hence,

$$
\begin{aligned}
P(F^r/E_1) &\;=\; 0.7 \\
P(F^r/E_2) &\;=\; 0.2
\end{aligned}
$$
and
$$P(F^r/E_3) \;=\; 0.1$$

Using the Bayes formula, the probability of getting a rise in the market if a rise report is given is:

$$P\left(E_1 / F^r\right) = \frac{P(E_1).P(F^r/E_1)}{P(E_1).P(F^r/E_1) + P(E_2).P(F^r/E_2) + P(E_3).P(F^r/E_3)}$$

$$= \frac{(0.6 \times 0.7)}{(0.6 \times 0.7) + (0.2 \times 0.3) + (0.1 \times 0.1)}$$

$$= \quad \textbf{0.854}$$

The market report changes the probabilities of each outcome and the table of Bayesian probabilities would look like this:

Market Outcome	Consultant's Prediction		
	Rise	Stable	Fall
Rise	0.85	0.38	0.32
Stable	0.12	0.56	0.32
Fall	0.02	0.06	0.37

So a new decision tree can be drawn. It is solved by tracing the expected outcomes back to the original objective. The bottom branch is the initial decision tree. However, the new tree cannot be solved until the probability of getting a rise, stable or fall report is known.

The probability of getting a rise report is given by the sum of the probabilities of getting a rise report under each outcome multiplied respectively by the probability of getting that outcome. Thus the probability of getting a rise report is:

$$= \quad (0.7)(0.6) + (0.2)(0.3) + (0.1)(0.1) \qquad = \quad \mathbf{0.49}$$

The probability of getting a stable report is .32 and the probability of getting a fall report is .19. Now that these values are substituted into the tree, the expected payoff with the consultant's report is:

$$= \quad (0.49)(219.6) + (0.32)(168.0) + (0.19)(141.1) \quad = \quad \mathbf{188.17}$$

Since the report costs £5,000, the net return would be £13,817. This is less than the outcome without the report and therefore the Pulsometer Pump Company will derive no major benefit from commissioning the report. This information can be shown on the new decision tree which incorporates the Bayesian analysis, **figure 4.9**.

Decision trees are a clear method of looking at the consequences of decisions. The problem is set out and the sequential decisions can be made. Obviously the accuracy of the method depends upon the accuracy of the expected outcomes and the probabilities. When further information becomes available, this can be incorporated with Bayesian analysis. The important aspects of decision trees are their logical structure and the revisions to probability that are possible.

Stochastic decision tree analysis

The stochastic decision tree approach, originally developed by Hespos and Strassman (1965) provides another method for analysing decision problems over time. It combines the logic of the decision tree analysis just discussed with the Monte Carlo simulation approach used in risk analysis. Risk evaluation typically treats investment decisions as if they were single-stage decision problems. However, it is more realistic to recognise that most decision problems have a large number of interrelated investment decisions over the time horizon of the project. Thus, if the two approaches can be combined, we have a potentially powerful technique for use in investment appraisal.

We can identify a number of features specific to the stochastic decision tree approach:

Figure 4.9   Decision tree with Bayesian theory

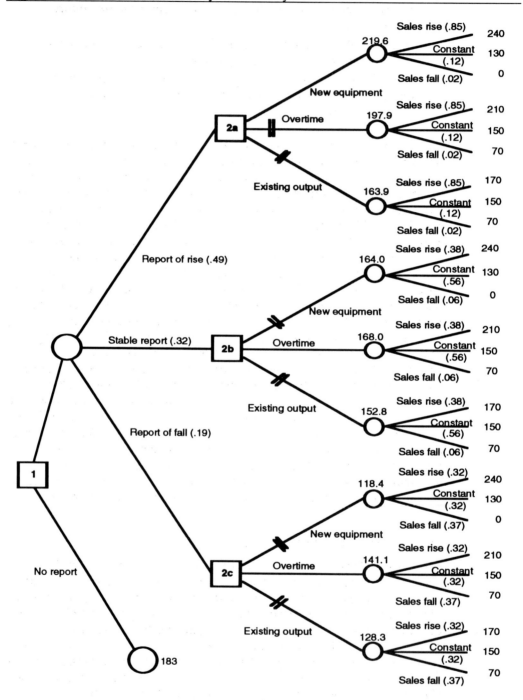

❏ instead of using the EMV principle, all factors, including chance events, can be represented by continuous, empirical probability distributions;

❏ the information about the results of decisions made at sequential points in time can be obtained in probabilistic form;

❏ the probability distribution of possible results from any particular combination of decisions can be analysed using the concepts of utility and risk.

The stages of stochastic decision tree procedure are similar to the ones outlined for the traditional decision tree approach. The decision-maker must search for the range of options, chance events which may affect him, and possible valuation problems prior to making the decision tree. Once a structure has been decided upon, the next stage involves evaluating all possible combinations of decisions in terms of probability distributions. Typically, the decision tree is used as a model for *forward-analysis*, and so each possible strategy path is evaluated in terms of a probability distribution based upon the NPV criterion. Such an approach has the particular advantage of being flexible. Once the probabilistic output has been determined, a modified rollback principle can be used. Branches of the tree would then be eliminated on the basis of stochastic dominance principles rather than the decision analysis norm of EMV. For example, a branch might be eliminated if it had a higher variance than an alternative branch. The dominance procedure will reduce the decision set and allow the analysis to focus on the more relevant decision paths.

When complex decisions are involved, we do not want to compound the problem by confusing the client with highly theoretical techniques. The strong advantage of the decision tree approach is that the decision tree is a visual and easily understood means of representing the sequential, multi-stage logic of a decision problem. Provided the client has access to information on such matters as quality and cost, then he will have much to gain by structuring his investment decisions in terms of a decision tree.

MULTI-ATTRIBUTE VALUE THEORY

In the last section, we saw how decision trees could be used to analyse decision problems that are sequential in character. However, just as there may be many options open to the decision-maker, so there may be a number of objectives. The client options may include which type of heating or air-conditioning system to install. His principal objectives may include minimising cost, risk, disruption, or maximising available space, visual appeal, user control, etc. A problem of analysis exists because the decision-maker's objectives may conflict. That is to say, improvement towards one objective may be achieved only at the expense of another objective.

Multi-attribute value theory is a formalised decision-making technique designed to counter the problems posed by conflicting objectives. In contrast to decision tree analysis, multi-attribute value theory provides a means of making more effective use of existing information, rather than identifying the need for additional information.

The decision-maker first attempts to produce the overall utility function, $U(x_1, x_2, x_3, ..., x_n)$, which reflects the relative desirability of an objective, having considered the relevant attributes $x_1, x_2, x_3, ..., x_n$.

Specify the utility function

Every alternative in making a decision can be described by a collection of attributes $x_1, x_2, ..., x_n$. These attributes should reflect the desired objectives of the decision-maker. Provided that the attributes can be represented by a scalar value or performance level, it should be possible to assess the combination of performance levels which most satisfies the decision-maker. The aim of multi-attribute theory is to obtain an overall utility function $U(x_1, x_2, ..., x_n)$ which yields a utility index or measure of worth for a given set of alternatives. A suitable utility function should capture the preference structure of a decision-maker. A decision-maker will therefore choose the alternative which maximises the function.

A useful simplifying assumption is that the overall utility function $U(x_1, x_2, ..., x_n)$ can be obtained from the utility functions of the individual attributes. That is:-

$$U(x_1, x_2, ..., x_n) = f\left(U_{(x_1)}, U_{(x_2)}, ..., U_{(x_n)}\right)$$

One example of this is the additive case.

$$U(x_1, x_2, ..., x_n) = \sum_{i=1}^{n} k_i v_i(x_i) \qquad (1)$$

A special form of utility function, the value function $v(x_1, x_2, ..., x_n)$, is assessed under conditions of certainty. Since some attributes may be known with certainty, it provides a convenient form. (For example, the value x_2 can be defined in terms of a discrete value rather than a probability range.) This avoids having to consider preference for hypothetical lotteries involving the attributes with no uncertainty, as would be the case with the conventional utility function. The additive relationship used in the following example is of the form:

$$v(x_1, x_2, ..., x_n) = \sum_{i=1}^{n} k_i v_i(x_i) \qquad (2)$$

Where $v_1(x_1)$ represents a single attribute value function (a single attribute value function describes the relationship between the performance level and the value to the decision-maker). Each scaling constant k_i, lies between 0 and 1, and the sum of all the scaling constraints equals unity.

Keeney and Raiffa (1976) have shown that such additive decompositions are appropriate when the condition of mutual preferential independence is satisfied. An attribute x_1 is preferentially independent of an attribute x_2 if the preferences for the specific outcome of x_1 are not dependent upon the values of x_2. If, correspondingly, the preference for x_2 is not dependent upon x_1, they are described as having mutual preference independence.

Case study

The selection of an air-conditioning system is used as a case study to illustrate the application of multi-attribute value theory using the above form (2). Since the preference expressed is problem specific, the eventual ranking applies only to the particular problem in hand. The stages in the analysis are as follows:

- ❏ generate a feasible set of options;
- ❏ specify suitable single attribute and joint utility functions;
- ❏ estimate the performance of each alternative including uncertainty. These include both objective and subjective assessments obtained from the appropriate specialists;
- ❏ evaluate the strategies using an overall utility function and applying sensitivity analysis to identify potential changes in ranking order.

Step 1   Generation of options

A set of ten air-conditioning systems was considered initially as shown in **table 4 A**. In this example, the application was a commercial office facility, and constraints concerning noise level and air distribution characteristics were imposed. Two alternatives were then rejected because of poor distribution characteristics (nos. 8 and 9). Option 10 was also rejected because of anticipated excessive noise levels.

Step 2   Specification of evaluation measures

Ten attributes were considered to have a significant influence on the viability of an air-conditioning system. These are listed in **table 4 B** along with the objectives and the type of evaluator used.

The life cycle costs include initial capital, operating and maintenance costs. Since these maintenance costs did not include the cost of disruption to occupants, another attribute x_3 was used to incorporate this loss of functional use.

Table 4 A   The alternative options for air-conditioning

1   Double duct overhead VAV for heating and cooling.
2   Overhead variable volume cooling and perimeter hot water heating.
3   Overhead variable volume cooling with terminal re-heat on perimeter for heating.
4   Perimeter four-pipe induction overhead for heating and cooling with internal VAV cooling.
5   Perimeter induction floor mounted for heating and cooling with internal zone VAV cooling.
6   Four-pipe fan coil overhead with ducted fresh air for heating and cooling.
7   Perimeter four-pipe fan coil wall mounted, with ducted fresh air for heating and cooling.
8   * Self-contained units air cooled, wall mounted.
9   * Self-contained water source heat pump, wall mounted with ducted fresh air.
10   * In floor VAV heating and cooling.

   * represents non-feasible alternatives

Table 4 B   Attributes for evaluating air-conditioning systems

Attribute	Objective	Evaluator	
x_1	Life cycle cost	minimise	money
x_2	Floor space encroachment	minimise	metres
x_3	Disruption to occupants during maintenance	minimise	subjective
x_4	Ceiling space requirement	minimise	metres
x_5	Perimeter partition flexibility	maximise	subjective
x_6	Module integration	maximise	subjective
x_7	Humidity control	maximise	subjective
x_8	User satisfaction	maximise	subjective
x_9	Degree of compatibility with potential expanded load requirements	maximise	subjective
x_{10}	Vendor viability and the availability of support	maximise	subjective

Step 3   Assessment of single attribute value functions
The decision-maker must next consider the individual value functions of attributes x_1 to x_{10}. This is achieved by examining the intervals or changes within a scale of measurement. The intervals must be rank ordered according to preferability. In other words the decision-maker must identify the interval of change for which they are prepared to pay the largest amount of money, the second largest amount, and so on.

Having rank ordered the intervals, the decision-maker should then indicate how many times as much money he is prepared to spend to increase the performance of each interval compared to the least valued interval change.

To illustrate this, consider attribute x_8, user satisfaction, with a performance range spanning -1 to +2. The decision-maker selected the interval -1 to 0 as the change for which he/she would pay the most money. In addition, the interval change 0 to 1 was perceived as being more important than interval 1 to 2 which was the least important interval. It was also stated that the decision-maker would be prepared to pay three times as much money for the transition from -1 to 0 compared to the interval 1 to 2. It follows from these statements that:

$$v_8(0) - v_8(-1) = 3 * \{v_8(2) - v_8(1)\} \tag{a}$$

$$v_8(1) - v_8(0) = 2 * \{v_8(2) - v_8(1)\} \tag{b}$$

Where $v_n(i)$ is the value of attribute n at point *i* normalised between the interval 0 to 1.

Moreover $v_8(-1) = 0$ since the least desirable level has a value of 0; the most desirable level has a value of 1 (i.e. $v_8(2) = 1$). This leaves us with two equations (a and b) and two unknowns ($v_8(0)$ and $v_8(1)$) which can be obtained by means of simultaneous equations.

Hence: $v_8(1) = 5/6$ and $v_8(0) = 1/2$.

The value function can then be represented graphically using the four known points (see **figure 4.10**).

It was assumed in this example that the overall expenditure of the company was sufficiently large that the preferences over cost were linear. The value function for x_1 was:

$$v_1(x_1) = (x^* - x_1)/(x^* - x_*)$$

Where $x_* = $ the lowest life cycle cost
$x^* = $ the highest life cycle cost.

Figure 4.10 The single attribute value function for attribute x_8

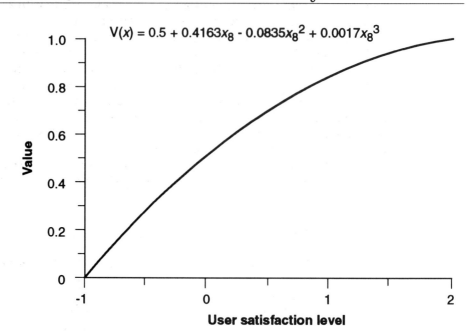

$$V(x) = 0.5 + 0.4163x_8 - 0.0835x_8{}^2 + 0.0017x_8{}^3$$

(y-axis: Value; x-axis: User satisfaction level)

This assumption was also used for attribute x_4, ceiling height.

Step 4   Scaling constant assessment

Trade-offs between competing objectives are assessed at this stage. The decision-maker must rank the evaluation measures in order of importance by considering the problem:

'Given that all the attributes are at their least desirable level, choose the attribute that you would most prefer to move from its least desired to its most desired level.'

This problem is repeatedly addressed, removing the chosen attributes each time until a ranking order of importance is achieved. In this example the ranking order was such that:

$$k_1 > k_6 > k_8 > k_{10} > k_9 > k_3 > k_2 > k_4 > k_5 > k_7$$

Having ranked the trade-off values, specific values for each scaling constant must then be obtained. This is achieved by pair-wise assessment of the attributes. The decision-maker must attempt to:

'Find a level for the more important measure that was equally preferable to the most desired level of the less important evaluation measure.'

In other words, define a jump in performance from its lowest level for the more important evaluation measure required to equate with a complete jump across the scale for the less important one.

To illustrate this step, consider the pair-wise evaluation using attribute x_4 ceiling space required, and x_9, load flexibility. The decision-maker has already decided that x_9 is the preferred evaluation measure to x_4 It is then necessary to establish an intermediate level in x_9 so that he/she is indifferent to x9 moving from its least desirable level to the intermediate level, or moving x_4 from its least desirable level to its most desirable level. The chosen intermediate level for x9 in this situation was 0.8 (i.e. $k_9 v_9(0.8) = k_4 v_4(0) = k_4$ since $v_4(0) =1$).

Table 4 C shows how the scaling constants were calculated substituting other constants for k_1 after determining the pair-wise indifference values.

Table 4 C   Solving for scaling constant a

k_n	Relationship to k_1	k_n equals*	Value of k_n
k_1	k_1	0.25	
k_6	$k_6 = k_1 v_1(2.9)$	$0.70k_1$	0.17
k_8	$k_8 = k_1 v_1(3.3)$	$0.57k_1$	0.14
k_9	$k_9 = k_1 v_1(3.6)$	$0.47k_1$	0.12
k_{10}	$k_{10} = k_9 v_9(1.5) = k_1 v_1(3.6)v_9(1.5)$	$0.38k_1$	0.09
k_3	$k_3 = k_9 v_9(1) = k_1 v_1(3.6)v_9(1)$	$0.27k_1$	0.07
k_2	$k_2 = k_9 v_9(1) = k_1 v_1(3.6)v_9(1)$	$0.27k_1$	0.07
k_4	$k_4 = k_9 v_9(0.8) = k_1 v_1(3.6)v_9(0.8)$	$0.22k_1$	0.05
k_5	$k_5 = k_4 v_4(150) = k_9 v_9(0.8)v_4(350)$ $= k_1 v_1(3.6)v_9(0.8)v_4(350)$	$0.09k_1$	0.02
k_7	$k_7 = k_4 v_4(200) = k_9 v_9(0.8)v_4(200)$ $= k_1 v_1(3.6)v_9(0.8)v_4(200)$	$0.05k_1$	0.01

$$\Sigma = 4.02\ k_1 \quad \Sigma = 1.00$$

*   Solving $\Sigma k_1= 4.02k_1 = 1$ yields $k_1 = 0.249$, from which the other ks are evaluated.

Step 5   Estimating the performance of options
Performance levels for each of the seven feasible options were estimated. These are listed in **table 4 D**. Life cycle costs, x_1, represent an uncertain variable, and it is therefore necessary to consider the probability distribution. The probability distribution may be approximated using a three-point estimate, employing the Pearson-Tukey approximation:

$$
\begin{aligned}
v_1 &= x_1(0.05) & p &= 0.185 \\
v_m &= x_1(0.50) & p &= 0.630
\end{aligned}
$$

$$v_h \quad = \quad x_1(0.95) \qquad p \quad = \quad 0.185$$

Where $x_1(0.05)$ is the low estimate at the 5 percentile level of x_1, $x_1(0.50)$ is the median value at the 50 percentile level, and $x_1(0.95)$ is the high estimate 95 percentile value.

Table 4 D   Results of evaluation measure estimates

Measure	Description	Strategy 1	2	3	4	5	6	7
x_1	Life cycle	3.0	3.8	4.8	2.9	3.2	2.6	2.8
x_2	Floor space	0	75	0	0	300	0	300
x_3	Maintenance	0	0	-1	-2	-2	-2	-2
x_4	Ceiling space	600	600	600	600	450	400	400
x_5	Partition flexibility	2	2	2	1	1	0	1
x_6	Module integration	2	2	2	0	1	1	0
x_7	Humidity control	2	2	2	1	1	0	1
x_8	User satisfaction	2	1	2	0	2	0	-1
x_9	Load flexibility	2	2	2	0	0	1	1
x_{10}	Vendor viability	1	2	1	1	2	0	1

*   Median value of life cycle costs £x_{10}^6

Step 6   Evaluating strategies

The final evaluation stage involves calculation of the overall expected utility value. Since attribute x_1 is an uncertain variable, it is necessary to calculate the overall utility value for each of the three-point estimates to arrive at an expected utility value. That is:

$$U[x|S] = \Sigma u(x_1, x_2, x_3, ..., x_{10}) \, p(x_1|S)x_1$$

where $U(x_1, x_2, ..., x_{10})$ = utility function
$p(x_1 | S)$ = probability distribution of x_1 conditional on S being the selected strategy.
x_1 = set of possible values for evaluation measure x_1 (approximated to three discrete values).

Sensitivity analysis is subsequently used to assess the effects of small changes in the scaling factors. If the position of the highest ranking option remains during such an analysis, then a superior alternative will have been identified. **Table 4 E** shows the eventual ranking order of the seven feasible air-conditioning options.

Table 4 E   Ranking order of feasible alternatives

Rank	Alternative	Expected Utility
1	1	0.86
2	2	0.78
3	3	0.66
4	5	0.61
5	6	0.59
6	4	0.44
7	7	0.43

Summary

Decisions involving new technologies in facilities management require careful analysis. In order to make life cycle costing a viable tool in such an issue, it may be necessary to incorporate it into a multiple objective analysis similar to that discussed above. The ability to handle non-linearity (returns to scale) makes it a more defensible approach compared to conventional weighted evaluation matrices. Since Mechanical and Electrical Services are intimately linked to operations at the work-place, consideration of functionality is of paramount importance and, as such, should be explicitly incorporated into the decision framework. Moreover, such an analysis should reflect not only economic viability but resistance to technological obsolescence.

SENSITIVITY ANALYSIS

Sensitivity analysis is discussed in detail in chapter 6; the purpose in this section is to give a brief overview. The discussion relates to the use of sensitivity analysis for life cycle costing but the approach is applicable to a wide range of activities.

Sensitivity analysis is used to identify the impact on the total of a change in a single risky variable. The major advantage of sensitivity analysis is that it explicitly shows the robustness of the ranking of alternative projects. Sensitivity analysis identifies the point at which a given variation in the expected value of a cost parameter changes a decision. For example, when considering the life cycle costs, if the total costs of fuel exceed expectations by 10% does this change the preference between two alternative projects?

Sensitivity analysis is an interactive process which tells you what effects changes in a cost will have on the life cycle cost. By identifying the relative importance of risky cost variables, the decision-maker can adjust projects to reduce the risks and consider responses should the outcomes occur.

A spider diagram is an effective way of using sensitivity analysis. The steps are:

i)   Calculate the expected total life cycle cost by using expected values.

ii)   Identify the variables subject to risk using a decision tree approach.

iii)   Select one risky variable, which we can call 'parameter 1', and re-calculate the total life cycle cost using different assumptions about the value of this parameter. The life cycle chosen is recalculated assuming that the cost parameter changes by 1%, 2%, and so on.

iv)   Plot the resulting life cycle costs on the spider diagram, interpolating between the values. This generates the line labelled 'parameter 1' as shown in **figure 4.11.**

v)   Repeat stages iii) and iv) for the other risky variables.

Each parameter line on the spider diagram indicates the impact on the life cycle costs of varying the value attributed to a particular parameter within the defined range. The flatter the line, the more sensitive will be the life cycle costs to changes in that parameter. In **figure 4.11**, total life cycle is much more sensitive to variation in parameter 1 than it is to variation in parameter 2.

Spider diagram

The spider diagram tends to appear more difficult to read when more variables are plotted. The practical answer is to have several spider diagrams. We would recommend having a spider diagram for the financial and capital aspects of the project, and a separate spider diagram for running costs.

The next question arises is whether there is likely to be a linear relationship between percentage changes in costs and changes in the expected value for total life cycle costs. In general, the spider diagram lines will not be linear, since if a running cost increases by x per cent it will be a relatively larger component of overall life cycle costs. Moreover, individual cost parameters may vary in many different ways.

Figure 4.11 Non-linear spider diagram

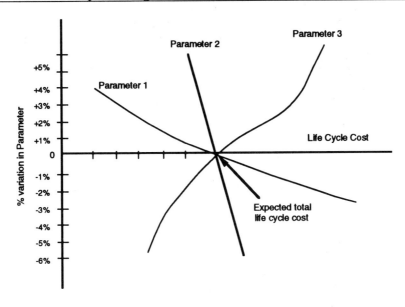

Figure 4.12 Severn Barrage project

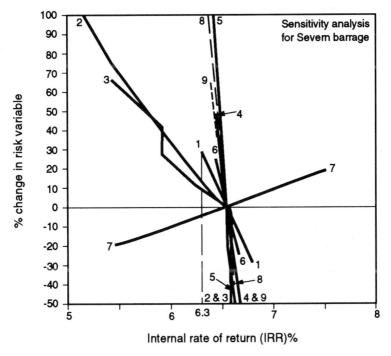

Source: *Hayes and Perry*

Hayes and Perry (1987) found estimates of sensitivity for the Severn Barrage project that were kinked: a feature that will characterise, in particular, intermittent costs, or parameters such as replacement periods for selected items of equipment. It is only the cost parameter in equation (3) which is linear in our theoretical example, because the costs are aggregated to one numerical total.

The discussion of sensitivity analysis so far has not involved the element of risk. Sensitivity analysis shows the effect of a change to a risky variable but it would be useful to know how likely it is that the cost parameter will vary within a particular range. This will require some form of probability analysis. In particular, we can use the central limit theorem to develop a tighter description of likely variability in costs. For each risky parameter, objective and subjective information can be used to identify the parameter's variance. This also gives us an estimate of its standard deviation. We then know from the central limit theorem that there is a 95% probability that the parameter will lie within ±2 standard deviation of its expected value, the value used in the original calculations. This can be used to define a 95% confidence probability contour as in **figure 4.13**. Point A in this diagram, for example, is defined as follows: there is a 95% probability that parameter A will lie within $\pm a_1$% of its expected value.

Figure 4.13 Probability contours

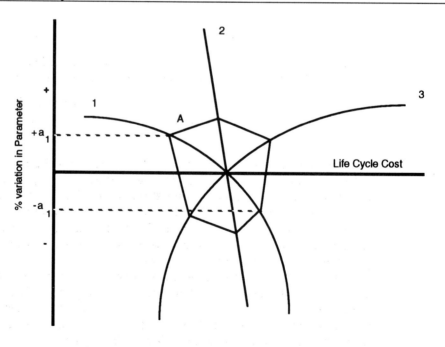

Sensitivity analysis provides clues to further investigation. It provides the critical factors for any forecast. For example, if the project is very sensitive to demand, then it will be worthwhile to spend resources establishing better demand forecasts. The sensitivity analysis shows the value of information regarding variables for a project, which of the parameters should be considered, and which it might be expedient to ignore. Evidence has shown that the consideration of the eight largest risks will typically cover up to 90% of the total (Porter, 1981). Consequently, the identification of these risks by sensitivity analysis can yield savings in information requirements and costs.

MONTE CARLO SIMULATION

> **Simulation is the art and science of designing a model which behaves in the same way as a real system. The model is used to determine how the system reacts to different inputs.**

Simulation is a word which is in common use today. The term simulation describes a wealth of varied and useful techniques, all connected with the mimicking of the rules of a model of some kind.

> *Simulate,* **verb transitive. Feign, pretend to have or feel, put on, pretend to be, act like, resemble, wear the guise of, mimic.**
>
> **(The Concise Oxford Dictionary)**

The most commonly known type of simulation is a flight simulator used to train pilots. Flight simulators can introduce all types of hazards to the pilot by simulating the live situation. Simulation techniques are used extensively in industry, for instance, in construction; different weather patterns can be simulated to determine the impact they will have on the construction schedule. Equally, simulation techniques can be used in the evaluation of cost for construction projects.

Simulation is a further method of analysing risk; it is basically a means of statistical experiment. Monte Carlo analysis is a form of stochastic simulation. It is called Monte Carlo because it makes use of random numbers to select outcomes, rather as a ball on a roulette wheel stops, theoretically at random, to select a winning number.

101

The Monte Carlo simulation will require sets of random numbers to be generated for use in testing various options. Random numbers could be selected in a variety of ways such as picking a number out of a hat, or throwing a dice. In reality, using a computer program is the most effective method of generating sets of random numbers. Monte Carlo simulation is described in detail in chapter 8; hence only a brief description is given here.

Figure 4.14 A cumulative frequency curve

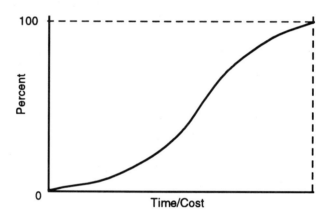

Simulation makes the assumption that parameters subject to uncertainty can be described by probability distributions. In Monte Carlo simulation a large number of hypothetical projects are generated to reflect the characteristics of the actual project. Each simulation (or iteration, as it is known) is accomplished by replacing a risky variable with a random number drawn from the probability distribution used to describe that variable. The way the distribution mimics uncertainties is discussed in detail in chapter 7. It is usual to make at least 100 iterations which then build up into a frequency distribution for the whole project. Statistical methods can then be used to calculate confidence intervals and so on. Cumulative frequency curves are also usually presented as part of the results. From these it is a simple matter to read off the likelihood that a certain activity will not exceed a given time.

PORTFOLIO THEORY

The aim of portfolio analysis is to maintain the expected return and to minimise the risks of the portfolio. An investor holds a portfolio which consists of various types of investments such as gilts, stocks and properties. Individual investments in a portfolio have their own return and risks. In portfolio theory, while the total return is a weighted average of individual investments, the overall risk of the portfolio cannot be measured in the same way. A correlation coefficient indicates the relationship between each of the individual investment risks. The coefficient is within a range between +1 and -1; a correlation coefficient of +1 indicates that if the risk of an investment increases by one unit then the overall risk of the portfolio will increase by one unit. Zero correlation coefficient means there is no overall correlation between individual investments in a portfolio. The overall risk of the portfolio can be diversified if the correlation coefficient is less than +1.

It is not possible to totally eliminate the risks of a portfolio. There are two kinds of risk: specific risks which are associated only with a particular investment in the portfolio, and market risks which affect all individual investments. Market risks such as the economic climate or taxation policy cannot be diversified and hence the overall risk of the portfolio can never be zero.

The application of portfolio analysis in the construction industry

Building firms can create a project portfolio, in which each project has its own expected return and risk. The overall risk of the portfolio will be reduced because no two buildings are the same, that is, the correlation coefficient is less than +1.

Figure 4.15 Risk reduction by diversification

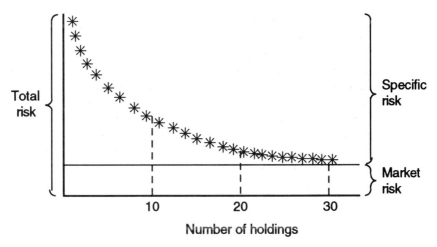

Housing and high technology commercial offices provide a useful illustration. Housing construction for public or private clients is less risky because it tends to use traditional materials and design. Since housing is relatively traditional and can be small scale with few barriers to entry into the market, tenders will be more competitive and the return likely to be lower. On the other hand, office construction has a greater risk because high technology materials may be used with innovative design, but not all contractors are capable of handling such projects and so the market is less competitive and the expected return should be higher. These two types of construction are not perfectly correlated to each other, so the risk in the portfolio of a contractor who undertakes both housing and office construction is diversified and a better combination of return and risk is achieved than is available from a concentrated portfolio.

Of course, such diversification is not without cost. In the investment world, an investor can maintain a balance between the return and risk of the portfolio by buying and selling assets. Such an active management method is not always applicable in the construction industry. Once a contractor enters into a contract, he has an obligation to complete the project.

Building in different sectors of the property market diversifies risk only at the expense of expertise. A construction firm is willing to build in all property sectors only if it has sufficient expertise. Otherwise, diversification will slow down the output of the construction team and hence increase project cost and decrease competitiveness. A trade-off is necessary between the gain that diversification gives through improving the risk/return mix, and the loss of economies of scale and learning economies reducing his exposure in any one project. The moral of the portfolio theory story is, 'Don't put all your eggs in one basket'!

STOCHASTIC DOMINANCE

Stochastic dominance is the analysis of probability distributions without specifying whether the mean or the variance are the parameters to be considered. The whole of the distribution is considered, and the only assumption is that 'more is preferred to less'. The criterion strongly orders alternative projects precisely, because of the lack of restrictive assumptions. It classifies projects that are efficient and those projects that are dominated by the set of efficient projects. Project F will first order stochastically dominate (FSD) G if the cumulative probability function of F is always below or equal to that of G.

We shall follow an example from Ward (1979). Two probability distributions are constructed from the following observed returns.

Portfolio F   1.8%, 2%, 6%, 7%
Portfolio G   1%, 3%, 3%, 6%

To simplify the example we assume that the probability distributions are constructed on the assumption that the chances of the observations occurring are all equal. It is also convenient to compare the distributions including all the rates of return. Thus portfolio F has the additional observations 1 and 3 although these are set at zero probability. These are portrayed in the following table.

Portfolio F

Return %	1.0	1.8	2.0	3.0	3.0	6.0	7.0
Probability $f(x)$	0.0	0.25	0.25	0.0	0.0	0.25	0.25

Portfolio G

Return %	1.0	1.8	2.0	3.0	3.0	6.0	7.0
Probability $g(x)$	0.25	0.0	0.0	0.25	0.25	0.25	0.0

First degree stochastic dominance is given by:

FSD = The probability function $f(x)$ is said to dominate the probability function $g(x)$ by FSD if and only if $F_1(x_n) < G_1(x_n)$ for all $n < N$ with strict inequality for at least one $n < N$ where

$$EUV = \sum [p_i U_i(X_i)] \qquad n = 2, 3, ..., N$$

In the example therefore, the difference between the cumulative probability distributions is calculated.

Cumulative distributions of illustrative portfolios

Return		1.00	1.80	2.00	3.00	3.00	6.00	7.00
Portfolio F $f(x)$		0.00	0.25	.25	0.0	0.0	0.25	0.25
$F_1(x)$		0.00	0.25	0.50	0.50	0.50	0.75	1.00
Portfolio G $g(x)$		0.25	0.00	0.00	0.25	0.25	0.25	0.00
$G_1(x)$		0.25	0.25	0.25	0.50	0.75	1.00	1.00
$F_1(x) - G_1(x)$		-0.25	0.0	+0.25	0.0	-0.25	-0.25	0.0

For F to first order stochastically dominate G the difference $F_1(x) - G_1(x)$ must always be negative or zero. Since at the observation $x = 2$ the difference is positive it cannot be concluded that F dominates G by FSD. This illustrates the problems surrounding stochastic dominance criteria.

Although stronger orderings can be produced, the outcome is that of a set of efficient portfolios or projects. It is the most theoretically rigorous criterion but in many cases cannot choose between alternative projects. Stochastic dominance tends to show the *efficient frontier*. Its applications have been fairly limited because it does not tell us how to choose between alternative projects that have different returns and different risks. It may be that in the future the criterion may be more widespread and certainly it can be used as a screening aid in comparisons of different cumulative distributions such as that used in Monte Carlo simulation.

Figure 4.16 Stochastic dominance

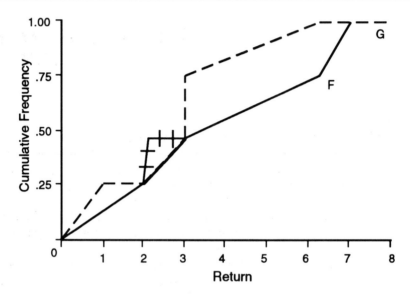

CONCLUSION

A number of decision-making techniques have been considered, in both a theoretical and technical way. These techniques are not exhaustive, but do offer some degree of assistance in the decision-making process. It is necessary to understand these techniques in order to fully appreciate and be able to interpret risk and uncertainty.

The conclusion to be drawn is that there is a wide range of tools for the performance of risk management. It will generally be the case that more powerful techniques are more sophisticated and more expensive in data and time requirements. What is necessary is to choose the most powerful tool consistent with constraints imposed by both data and time limits.

5

UTILITY AND RISK ATTITUDE

INTRODUCTION

$\mathbf{W}$e are all different and we all make decisions. Our attitude to risk is important yet decision-making techniques rarely take account of this. The decision-maker has to make a choice from multiple possibilities, any one of which will result in a pay off or outcome which cannot be known in advance. This situation is the 'risk environment'.

Different people will make different decisions within the same risk environment. Some people are risk takers by nature, willing to take additional risks on the expectation of a higher return. Others are risk neutral, indifferent to return except if it can be calculated to be worth the risk. Yet others are risk averse, willing to sacrifice the possibility of a higher return even for a relatively small risk.

In general, we all have preferences and these are of two types. Firstly, there is a **direct** preference for something. For instance, where a contractor must choose between two suppliers' quotations and he chooses A in preference to B because A has been more reliable with deliveries in the past. Secondly, there is an **attitude** preference. A contractor may wish to play safe and develop five town houses on a site he owns where there is a 90% chance of a profit of £50,000, in preference to developing the same site with two executive houses with a 60% chance of gaining £100,000 but a 40% chance of losing £20,000.

In practice, most monetary decisions involve a complicated mix of the above two factors above. Any theory of choice must explain both the direct preference and the attitude to risk. Much more attention has been devoted to the study of how people recognise and evaluate risks and how they make choices in the risk environment than to how they attempt to modify the risks they face. There are three steps to be considered:

Step 1   Evaluating the risk exposure
Step 2   Understanding the risk attitude

Step 3   Reaching a decision

RISK EXPOSURE

Risk exposure is concerned with the amount of risk a person or organisation is facing. Risk exposure can be measured by probability distributions which give a profile of the risk being encountered.

A company proposing to design, build and operate a high speed rail link between Durban and Johannesburg would be interested in the design and development cost, the construction cost, the operating cost and the traffic density and the income over time. The main questions concern the confidence that can be placed on the estimates of cost and income and the consequences for the company should the estimates be unreliable or inaccurate. If the project failed to meet the investment targets and the consequence of failure was insolvency of the company, then obviously the company should not be exposed to that amount of risk.

There are both informal and formal approaches that decision-makers might follow when incorporating risk attitudes into project evaluations. The informal approach is where the distribution profile, the mean and standard deviation, and other relevant statistics can be considered together with the risk exposure. Statistics are a tool that helps to measure the risk exposure, but the decision also has to be made on objective or intuitive perception based upon experience, knowledge and wisdom.

This informal approach to decision-making lacks any standard procedure for measuring personal or institutional risk attitude when making a choice. The world is littered with companies who have over exposed themselves to debt and risk.

UTILITY THEORY

A more formal approach to measuring the decision-maker's attitude towards risk uses utility theory.

Risk attitude is concerned with the trade-off that people will make between uncertain payoffs of known probability and sure payoffs, again with known probability. The trade-offs are determined by asking decision-makers to specify how much sure money (the certainty equivalent) must be received to make them indifferent between the certainty equivalent and the expected value of a given amount that is not certain. For instance, if a person was given a choice between a 50% chance of winning £10,000 on a roulette wheel and a 50% chance of winning nothing or a 100% chance of winning £1,000 betting on a horse race and for a £500 stake winning, there would be some point where the decision-maker would be indifferent between the roulette wheel gamble and the horse race gamble. The decision is coloured by how much the gambler can afford to lose and how much he needs to win. The relationship is between money and utility, where utility means the satisfaction the decision-maker receives from given quantities of money.

Although the expected utility theory was first formulated over 200 years ago, it is only comparatively recently that the deeper significance of this approach for decision-making has been recognised.

A significant breakthrough was made by a brilliant young British logician, Frank Ramsey, in a paper published posthumously in 1931, and later in 1944 by John von Neumann and Oskar Morgenstern in their work on the theory of games. Ramsey, and then von Neumann and Morgenstern wished to demonstrate the superiority of the expected utility hypothesis not only vis-a-vis expected monetary value but also with respect to other possible theories of behaviour. They achieved this by providing a justification for the use of expected utility to explain choices under conditions of uncertainty.

The utility theory says that when individuals are faced with uncertainty they make choices as if they are maximising a given criterion, the expected utility. The von Neumann Morgenstern utility, the cardinal utility or simply the utility function is a characteristic of an individual or a group - it is unique.

Expected utility is a measure of the individual's implicit value, or preference, for each policy in the risk environment. This measure is represented by a numerical value associated with each monetary gain and loss in order to indicate the utility of these monetary values to the decision-maker. The utility measure can also be assigned to outcomes that have no monetary value. For the moment, let us restrict ourselves to monetary payoff situations which are more straightforward.

The utility measures should be consistent, in order to reflect the preference of the decision-makers. The following rules must be obeyed:

❑ the more desirable an outcome, the higher the utility measure will be. For example, winning £50 without any risk will have a higher utility measure than winning £5 without any risk;

❑ if a decision-maker prefers outcome A to outcome B, and he prefers outcome B to outcome C, then A will be preferred to outcome C;

❑ if a decision-maker is indifferent between two outcomes, they have equal utility;

❑ in a situation involving risk, the expected utility of the decision equals the true utility of the decision. For example, assume that a particular strategy has an outcome O1 with a probability P1, and an outcome O2 with a probability P2 = 1 - P1. If we define the utility of O1 as U(O1) and the utility of O2 as U(O2), the expected utility of the strategy, which we define as the EU strategy is:

$$EU(strategy) = P1*U(O1) + (1 - P1)*U(O2)$$

EXPECTED MONETARY VALUE

Before we discuss the utility function, let us recall the method of Expected Monetary Value (EMV) for decision-making as dealt with in Chapter 4. The Expected Monetary Value (EMV) of each strategy is determined by multiplying the payoff of each outcome by its probability of occurrence and adding the products.

$$EMV = \Sigma[p_i M_i(X_i)]$$

For example, if an investor has two strategies, either hold £150,000 in cash or invest the £150,000 in a project for which the options of return are £300,000 with a probability of 0.5 and £0 with a probability of 0.5, the EMV of the investment return is:

$$EMV = 0.5 \times 300,000 + 0.5 \times 0 = £150,000$$

Under the probability choice criteria, the decision-maker's option is based on the rule:

$$Strategy\ option = MAX_i \{EMV_i\}$$

However, the two strategies have the same Expected Monetary Value (EMV) of £150,000, therefore, how can the investor make the decision of whether or not to invest.

The following example shows another implementation of the EMV. A construction company is hiring equipment with a value of £50,000. It can buy insurance for £500 which will pay for replacing the equipment if it is damaged. The probability of the equipment being damaged is 0.05.

There are two strategies:

S1: don't buy insurance   S2: buy insurance

and two events:

E1: equipment not damaged   E2: equipment damaged

Let us form the payoff table for this problem, taking losses as negative income. If the company buys the insurance, their cost will be £500, whether or not the equipment is damaged. The payoff matrix is shown along with the expected monetary value (EMV) computations.

Using EMV criterion, the company would select S1 and not buy the insurance. Yet, in practice, many construction companies would, and actually do, purchase insurance. The prospect of a £50,000 loss somehow outweighs the £500 payment even with the low probability of risk.

Probabilities   P(E1) = 0.95   P(E2) = 0.05

Payoff matrix

	E1: Not damaged	E2: Damaged	Expected Monetary Value
S 1 (don't buy insurance)	£0	- £50,000	EMV(S1) = 0.95 *0 - 50,000*0.05 = -2,500
S 2 (buy insurance)	- £500	- £500	EMV(S2) = -500 *0.95 - 500*0.05 = - 500

The above example is very simplistic, but it shows that the most crucial decision that must be analysed is the probability factor. If the factor is not realistic then the complete analysis will give the wrong information. Furthermore, people's decision-making in the risk environment is not always consistent with the theory of EMV. In other words, the EMV theory cannot properly explain people's behaviour in some risk situations. This is because the principal objection to using EMV maximisation as the chosen criterion for decision-making is that it ignores the attitude to risk.

Summary of the steps involved in calculating the expected monetary value (EMV) theory
1   Consider the various options available
2   Estimate the value of each option
3   Estimate the probability of each option ensuring there is realism in the probability factor selected
4   Multiply the value by the probability
5   Sum the EMV of each option
6   Select the biggest EMV among the options

THE UTILITY FUNCTION

To be of use in decision-making, utility values must be assigned to all possible outcomes because the decision-maker's choice will change according to the risk involved. This relationship between expected return and choice is commonly expressed by the utility function. It is illustrated in the form of a utility curve with the utility scale on the vertical axis and expected outcome on the horizontal axis.

In summary, a utility function has the following properties:

❏ each possible outcome is defined by a single number;
❏ the outcomes are ranked in order of preference;
❏ the objective is to maximise expected utility.

There are several methods applied to the assessment of a person's utility function, such as the von Neumann-Morgenstern (NM), the Ramsey method, and direct measurement. The NM method is considered to be the most effective and will be explained briefly here.

The first step in deriving a utility function by the NM method is to determine two monetary outcome values as reference points. For convenience, we will look at the most favourable and least favourable monetary outcomes in a decision situation. We then assign utility values to these two reference points. Since utility is an ordinal rather than a cardinal concept, these utility values are arbitrary. All that is necessary is that utility increases with the monetary gain. For convenience again, therefore, we might assign arbitrary utility values of 1 and zero, respectively, to these extreme monetary outcomes. Assume that the monetary return outcomes of a gamble range from £0 to £300. So we choose extreme monetary values of £0 and £300, assigning a zero utility to £0 and a utility of 1.0 to £300. That is,

$U(£0) = 0$ *and* $U(£300) = 1.0$

The second step of the NM method is to assign the utility values for all the other monetary outcomes lying between these two extreme monetary outcomes. The utility values are determined in the NM method as the basis of the concept of certainty equivalent.

Assume that the decision-maker has to choose between two strategies:

Strategy A   a given amount of money with certainty (certain money)
Strategy B   a risky environment with probability p of winning £300 and probability $(1-p)$ of winning £0.

To determine a certainty equivalent of strategies A and B, we can change the parameter values of either strategy A or B, or both, until a certainty equivalent is obtained. For convenience, assume $p = 0.5$ and $(1-p) = 0.5$ for strategy B and that the certain money of Strategy A increases from zero. When the certain money of strategy A reaches £100, it makes the decision-maker indifferent between strategies A and B (see **figure 5.1**). Therefore £100 is the certainty equivalent between strategies A and B. Its utility equates to the expected utility of strategy B:

$U(100) = pU(£300) + (1-p) U(£0) = 0.5*1 + 0.5*0 = 0.5$

Here we introduce two simple and effective methods to determine the utility values lying between the range based upon the concept of certainty equivalent.

Method 1
Determine $U(x_3)$ $(x_1<x_3<x_2)$ by using two known utilities $U(x_1)$, $U(x_2)$ and their probabilities p and $(1-p)$. Assuming $p = 0.5$ for strategy B, and its EMV = £150, the basic principle is as follows:

Offer the decision-maker a certain money x_3 of strategy A starting from x_1, until this certain money x_3 makes the decision-maker indifferent between strategies A and B. For example, say $x_3 = £100$, then:

$$U(£100) = 0.5*1 + 0.5*0 = 0.5$$

Figure 5.1   How to determine a certainty equivalent

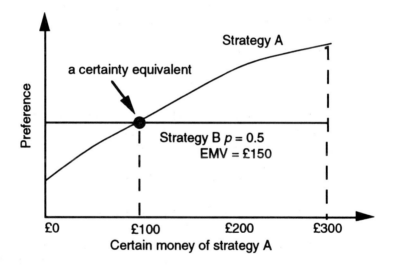

Certain money of strategy A

Next assume $p(£300) = 0.5$ and $p(£100) = 0.5$. Between £300 and £100 we ask the same question of the decision-maker. He is indifferent to certain money of £200 from strategy A and the payoff of £300 and £100 from strategy B. Thus,

$$U(£200) = 0.5*1 + 0.5*0.5 = 0.75$$

Then, assume $p(£100) = 0.5$ and $p(£0)$between £100 and £0. Following the same procedure above, the decision-maker is indifferent to £50, then:

$$U(50) = 0.5*0.5 +0.5*0 = 0.25$$

The utility function of the decision maker is shown in **figure 5.2.**

Method 2

The process of measuring a decision-maker's utility function under the given situation can be considered as shown in **table 5 A**. The decision-maker is asked to indicate for cell in each column whether he prefers A or B or is indifferent.

Table 5 A   Measurement of a decision-maker's utility function

		Probabilities that outcome of policy will be £300 or £0										
Certain	£300 (P)	0.0	0.1	0.2	0.3	0.4	0.5	0.6	0.7	0.8	0.9	1.0
Money(Policy A)	£0(1-P)	1.0	0.9	0.8	0.7	0.6	0.5	0.4	0.3	0.2	0.1	0.0
£310		A	A	A	A	A	A	A	A	A	A	A
£300		A	A	A	A	A	A	A	A	A	A	I
£280		A	A	A	A	A	A	A	A	A	A	B
£260		A	A	A	A	A	A	A	A	A	A	B
£240		A	A	A	A	A	A	A	A	A	I	B
£220		A	A	A	A	A	A	A	A	A	I	B
£200		A	A	A	A	A	A	A	A	A	B	B
£180		A	A	A	A	A	A	A	A	I	B	B
£160		A	A	A	A	A	A	A	A	I	B	B
£140		A	A	A	A	A	A	A	I	B	B	B
£120		A	A	A	A	A	A	A	I	B	B	B
£100		A	A	A	A	A	A	I	B	B	B	B
£80		A	A	A	A	A	I	B	B	B	B	B
£60		A	A	A	A	I	B	B	B	B	B	B
£40		A	A	A	I	B	B	B	B	B	B	B
£20		A	A	I	B	B	B	B	B	B	B	B
£10		A	I	B	B	B	B	B	B	B	B	B
£0		I	B	B	B	B	B	B	B	B	B	B
-£10		B	B	B	B	B	B	B	B	B	B	B

Taking one column at a time, the entire table is completed. For example, start from the bottom of the first column of the table, and ask the decision-maker to choose between losing £10 (strategy A) or gambling (strategy B) with a probability 0 of winning £300 and a probability 1.0 of losing £0. Strategy B is obviously preferred. Moving up to another cell, seek a preference between receiving £60 (strategy A) or gambling (strategy B) with a probability 0.4 of winning £300 and a probability 0.6 of winning £0. The decision-maker may feel indifferent between these two options which is shown by I in the table.

Obviously, except for the first and last columns, we would expect different decision-makers to select a different pattern of indifference points. Their selection will depend on many personal characteristics, such as attitudes towards risk, company financial situation, and so on.

From **table 5 A**, we can calculate the utility values by using expected utility method:

$U(£0) = 0*1 + 1*0 = 0$ $U(£10) = 0.1*1 + 0.9*0 = 0.1$

$U(£20) = 0.2*1.0 + 0*0.8 = 0.2$ $U(£40) = 0.3*1.0 + 0*0.7 = 0.3$

$U(£60) = 0.4*1.0 + 0*0.6 = 0.4$ $U(£80) = 0.5*1.0 + 0*0.5 = 0.5$

$U(£140) = 0.7*1.0 + 0*0.3 = 0.7$ $U(£180) = 0.8*1 + 0*0.2 = 0.8$

$U(£240) = 0.9*1 + 0*0.1 = 0.9$ $U(£300) = 1.0*1 + 0*0 = 1.0$

For example, in completing the table, the decision-maker had an indifference point at £100 in the column where $p = 0.6$. This means the utility value of £100 was:

$$U(£100) = pU(£300) + (1-p)U(£0) = 0.6 \times 1 + 0.4 \times 0 = 0.6$$

Thus, there is an indifference point of (£100, 0.6). As this process is repeated, the points on the curve are produced.

The indifference points with p (or U) on the vertical axis and certain money on the horizontal axis can now be plotted. This produces a utility curve as in **figure 5.2**.

Figure 5.2 shows the decision-maker's utility curve for our example. The curve reflects his risk attitudes over the range of payoff from £300 to £0 under various uncertain situations.

One difficulty that has been identified by researchers is that many decision-makers found it difficult to distinguish between events with probabilities with values such as 0.05, 0.1, 0.2 and 0.8, 0.9 and 0.95. To them these events were either extremely likely or unlikely. The result will be a poorly determined upper and lower ends of the range.

To return to the above example, the utility function assigns a utility of 0.6 to a certain outcome of £100. This is equivalent to a gamble which has a payoff of £300 with probability 0.6 and £0 with probability 0.4 giving an expected monetary value of £180.

GENERAL TYPES AND CHARACTERISTICS OF UTILITY FUNCTIONS

We would like to be able to say something more specific about the shape of the utility curve. Utility curves can be divided into three broad categories as shown in **figure 5.3**, dependent on whether the decision-maker is a risk seeker, a risk avoider or risk neutral. Of course, these curves are by no means the only possible forms of utility functions.

Figure 5.2   The utility function of the case study

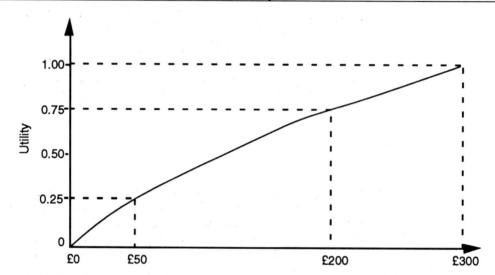

It could be, for example, that the decision-maker is a risk seeker over one range of monetary outcomes and a risk avoider over others. We shall not consider these mixed cases. The utility curve depicted in **figure 5.2** is consistent with the risk avoider, which implies that people ordinarily attach greater utility to a larger amount of money.

Curve A illustrates the utility curve of an individual who always takes certain money as the priority. He is a risk avoider.

Curve B represents the behaviour of a person who is neutral to risk. He is indifferent between certain and uncertain money.

Curve C shows the utility function for a risk seeker, who willingly accepts gambles that have a smaller expected value rather than an alternative payoff received with certainty.

In our example, if $x_1 = £300$ and $x_3 = £0$, then:

For the risk avoider $x_2 < £150$
For the risk neutral $x_2 = £150$

For the risk lover $x_2 > £150$

Figure 5.3   Three types of utility function curves

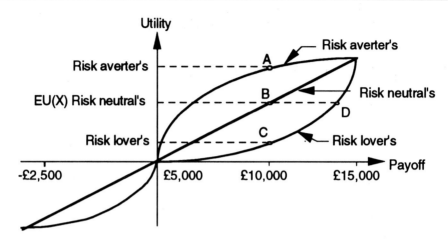

A person's utility function will not necessarily remain constant over time. As wealth and other factors in life change, the values will change, and consequently the utility function will change as well.

The risk curves shown in **figure 5.3** oversimplify the real situation. There has been considerable research into the psychology of preference judgements. For instance, a businessman's utility function is influenced greatly by his wealth. If the range of the businessman's utility function for assets includes wealth, then his utility function will be risk averse for normal business risks but very risk seeking near to bankruptcy.

Furthermore, there will probably be a different utility function for an individual compared with a group. Often it is necessary for groups of people to choose between competing options. Utility evaluations of individuals cannot simply be aggregated to obtain group utility, particularly as all the executives will have utility functions of different shapes. The most acceptable solution is for the top decision-making group, the directors or partners, to give guidance on their risk attitude.

Broadly then, utility theory is a decision theory concerned with choice from among a set of risky options. The theory suggests that choice of an optimal policy is determined by choosing the policy having the highest Expected Utility Value (EUV).

THE DIFFERENCE BETWEEN EUV AND EMV IN PRACTICE

Both EUV and EMV are decision criteria used in situations subject to risk. EMV is a common decision criterion used in practice, mainly because it is easy to understand and apply. However, the principal objection to using EMV maximization as the chosen criterion for decision is that it ignores attitude to risk. EMV and EUV can be guaranteed to give the same decision advice only if the decision-maker is risk neutral.

THE USE OF UTILITY THEORY IN CONSTRUCTION

Basic principle for the application of the theory

Many individuals in the construction industry will be sceptical about the formal concept of utility because they will distrust the use of graphs to replace judgement, in the same way that computers were mistrusted in their early days. The utility function is not a substitute for expert judgement, it is an aid - a decision-making tool.

The following example illustrates the implementation of the utility concept in construction by two companies, one small and one large.

There are two contracts K and L. The firms cannot undertake both, and to simplify the illustration, it has been assumed that each project can lead to three possible outcomes. The probabilities and payoffs are shown in **table 5 B.**

Table 5 B   Probabilities and payoffs of contracts K and L

Contract K		Contract L	
Probability	Outcome	Probability	Outcome
0.2	£100,000	0.2	£200,000
0.5	£500,000	0.7	+£150,000
0.3	- £10,000	0.1	-£0

Under the EMV approach, the appropriate calculations are as follows:

$$EMV(K) = £100,000 * 0.2 + £500,000 * 0.5 - £10,000 * 0.3 = £447,000$$
$$EMV(L) = £200,000 * 0.2 + £150,000 * 0.7 + £0 * 0.1 = £145,000$$

Expected utility value of the large company is:

Contract K   Contract L
0.2 * 1.0 + 0.5 * 0.8 + 0.3 * 0 = 0.6   0.2 * 0.5 + 0.7 * 0.3 + 0.1 * 0 = 0.31
Expected utility value of the small company is:

Contract K   Contract L
0.2 * 1.0 + 0.5 * 0.9 + 0.3 * 0 = 0.65   0.2 * 0.85 + 0.7 * 0.8 + 0.1 * 0 = 0.73

Table 5 C   The utility values of each company under the different outcomes

Outcomes	Utility value for the large company	Utility value for the small company
£100,000	1.0	1.0
£500,000	0.8	0.9
£200,000	0.5	0.85
£150,000	0.3	0.8
£0	0.0	0.0
-£10,000	0.0	0.0

Based on these calculations:

The large company would choose Contract K.
The small company would choose Contract L.

6

RISKS AND THE CONSTRUCTION PROJECT - MONEY, TIME AND TECHNICAL RISKS

INTRODUCTION

This chapter identifies the risks on any construction project, divided, for simplicity, into money, time and technical risks. Money, of course, encompasses the whole process, being the driving force of the project.

As a project moves from the investment and development appraisal stage through to occupancy so each phase develops and more information becomes available. Even at the very earliest stage, it is likely that the investment surveyor, architect, engineer and quantity surveyor will all be providing an input on the feasibility of the scheme from a design, cost and value viewpoint.

The money, time, and delivery sequences, discussed below, are intrinsically linked. If the project is not completed on time, somebody, whether client or contractor, is likely to suffer financial loss and expense.

MONEY AND DELIVERY SEQUENCE

Figure 6.1 shows the money delivery sequence which is categorised into three phases:

- **Phase 1 - Investment and development appraisal sequence** where the income and value are balanced against the cost. The private sector client will evaluate all the income and expenditure options. For instance, with a proposed new department store the initial capital cost and the market value of the investment will be balanced against the sales revenue and the direct cost of sales and the store operating and ownership costs. The public sector also makes these judgements, albeit on a different basis. The proposal for a new old people's home will be based not upon sales revenue but upon competing needs and how the home can be financed. The public sector has limited capital and competing priorities with an almost insatiable demand for new buildings and facilities.

- **Phase 2 - Design and construction sequence** from budget allocation through to completion of the project that will typically be greater than the initial capital costs.

- **Phase 3 - The occupancy sequence** where the expectations of the sales, or rental income revenue estimates for the private sector client, will be proved correct or otherwise. This phase can last from one to fifty years or so, and each year the facility will need to be maintained, heated, cooled, cleaned, insured and so on. In the past, the running costs have not been considered to any great extent at the design stage, yet the building will incur operations and maintenance costs throughout its life.

Investment and development sequence

In Chapter 1 the types of private sector clients of the building industry were characterised into three groups: ownership, investment, and property dealing. All three groups will have widely differing objectives, some will be seeking a long term investment, while others may view the project as a short term trading option. Investors have a bewildering choice of investment options; the choices range from risk free investment in long term treasury bonds to high risk investment in the futures market.

Markets are volatile and property belongs to a risk category which does not provide any guarantee of return, but there are four principal variables which underlie investors' decisions:

- ❑ returns that could be expected from other investments;
- ❑ the growth or decline expectation from the property market value and the rental income;
- ❑ the time horizon of the investment;
- ❑ the risk: intuitively, high risk projects should show a high rate of return, and low risk projects a lower rate of return.

Risk evaluation is tempered by the possibility of actual loss of capital or of loss of income, and the risk of irregularity of income, for example where a company is unable to regularly declare a dividend through lack of profitability.

Figure 6.1   Money (cost/revenue/value) delivery sequence

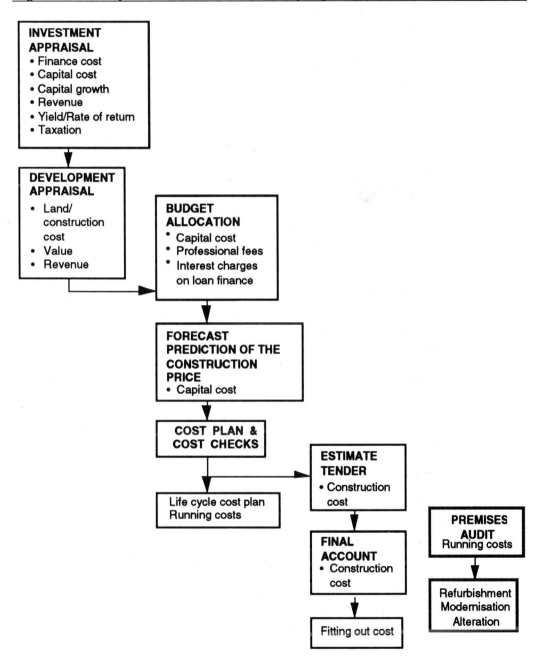

Property, as an investment, goes through phases. In boom times property is an attractive option because of the capital growth, which, in the 1970s and 1980s, was nearly always upwards. The unthinkable has happened in the 1990s and property prices have dropped in value. Reductions in value of 50% have been recorded on some sites; this hasn't happened in the past 50 years. A new era has arrived for property investment. Many large property and construction companies are in financial trouble because of writing down the value of their assets and the banks and financial institutions have become exposed to companies not paying their debts. Canary Wharf in London's Docklands was always a high risk project, but the scale of the potential short term loss was never perceived.

The financial design of a project has become as important as the architectural and engineering design. Financial design involves a feasibility study, investment appraisal, market study, and the options for financing the project. Not too long ago, financial appraisals could be done very quickly and simply. Recently, discounted cash flow techniques reflecting the time value of money have become widely accepted. These are useful because they adjust periodic cash flows according to a time weighted formula. As money has a time value, the timing of expenditure and the receipt of income are material considerations in the assessment of any investment. The net present value and the internal rate of return have become vital tools for investment appraisal.

The options for financing the project should be kept separate from the investment proposal unless they are inextricably linked. Providers of the company's capital come from financial markets made up of equity investors and debt holders. The company's task is to maximise the returns for those investors from its projects. The owner occupier wants a building to fulfil a function; to this end, he might use his own capital, borrowed capital, or he might issue shares if he owns a public company; there are an endless number of permutations to secure capital. Generally, the building will be wanted within budget, completed within the specified time, and to the desired quality. The investment appraisal must consider both the costs and revenues, and the value and rental growth and yield associated with the project.

Tables 6 A, 6 B and **6 C** show the considerations that the clients will take into account in a property investment decision, balancing the risk of and the reward from the investment. Most items in the tables are discussed elsewhere; hence discussion here is restricted to only a few of the issues.

The tables have been divided into:

- ❑ *cost* - to reflect the initial capital and ongoing long term costs;
- ❑ *operational/revenue* - to show the revenue earning and operational considerations when the building is in use;
- ❑ *value* - to reflect the issues associated with the value.

Cost considerations

The items in **table 6 A** are applicable to a commercial development in which the owner will be letting the building, whilst the tenant is generally responsible for fitting out and furniture and fittings. The increasing complexity of the mechanical and electrical engineering services means that the owner incurs costs to adequately balance the systems and to modify items to meet the tenant's requirements.

In view of rapid technological changes and evolving tenant requirements, there is increasing concern for both owners and users to be aware of the refurbishment, retrofitting, alteration, and modernisation costs. Buildings, like cars, are subject to deterioration and obsolescence; the mechanical and electrical systems are particularly subject to technological change.

Deterioration, as distinct from obsolescence, arises from wear and tear. Generally, the cost of maintaining a building system at a given condition rises steadily with age until a point is reached at which the cost of maintenance becomes prohibitive, and the end of the building's physical life is reached.

Table 6 A   Cost considerations

- Land purchase cost
- Adviser's fees on land acquisition
- Legal fees on land purchase
- Stamp duty on land purchase (taxation)
- Design Team fees
- Local government planning and building control fees
- Construction cost
- Interest charges on land and construction cost if loan finance
- Fitting out costs and commissioning charges
- Furniture and fittings
- Letting fees (if appropriate)
- Operating and maintenance costs (running costs)
- Modernisation and refurbishment costs caused by deterioration, obsolescence and market requirements
- Opportunity cost of money

However, in a commercial property there is a further factor; periodic refurbishment is required if the investment is to retain its initial attractiveness as a lettable property. Fashions, working patterns, changing standards, and technological advances all conspire to reduce the attractiveness of a property over time to both current and potential future tenants. The extent of refurbishment and modernisation depends upon how frequently such expenditure is envisaged. The longer the time periods, the greater the cost involved in order to restore the property to an acceptable standard.

Operational/revenue considerations

Obsolescence is irregular in nature and affects all building systems of the same type at the same time, thus involving risk and uncertainty. It is harder to control, being influenced by future events; it produces a relative loss of utility in contrast to deterioration which produces an absolute loss of utility. The building system may be as good as new, but made totally obsolete by a new and superior system.

Inflation will have an impact upon cash flows. For example, building costs will be affected by the impact of inflation, but rental levels will often rise at a different rate of inflation. Where rental levels are adjusted annually, the revenue should keep pace with inflation, but there is an added complication when there are periodic rent reviews, say every five years.

Table 6 B   Operational/revenue considerations

- Annual rental income
- Number of years between open market rent reviews
- Rental value growth rate
- Number of years between refurbishment /modernisation /alteration
- Time between site acquisition and revenue earning
- Service changes that can be recouped from tenant
- Taxation (Capital Gains Tax, Corporation Tax, Stamp Duty, Value Added Tax, Local taxes)

Owners will be seeking to balance the rate of return with the growth. For instance, if the rate of return on a property, allowing for risk, is 10% and the growth rate is expected to be 6%, a rack rented freehold benefiting from annual rent reviews, will give a yield of 4%. In other words, an annual income capitalised at 4% would show a rate of return of 10% per annum assuming the capital value increased by 6%.

The influence of taxation

The influence of taxation on an investor's decision cannot be ignored. The tax status of an investor will affect the investment decision. Given a discriminatory fiscal system which taxes income more heavily than capital gains, it is evident that investors with a high income tax liability will prefer those properties which offer low income yields but high capital gains.

Pension funds and life insurance companies are major investors in property. The taxation legislation is complicated but, in general, pension funds are not liable to income or corporation tax, and life insurance companies are taxed on their investment income rather than their trading profits. Hence, the tax status of investors has to be considered.

The tax position will also influence the way a project is financed. A project can be financed partly by debt and partly by equity. However, one of the principal advantages of utilising debt to finance a project is that interest payments are tax deductible. Therefore, the tax position should be considered at the development appraisal stage. There are two ways of doing this:

❑ by using a discount rate which is adjusted to the net of tax weighted average cost of capital. This procedure is not easy because the debt finance for the project must be the same as that employed by the firm;

❑ by calculating the net present value, assuming the project is equity financed, and then adding to this the present value of the tax break arising through the use of debt finance discounted at the cost of the finance.

In the case of a pension fund where no tax is paid, projects should be assessed as if they were wholly equity financed.

Both market value and investment value are needed when deciding upon a course of action. Market value represents the typical or most likely selling price of the property, and provides a point at which to begin negotiations between an owner and a potential buyer. Investment value, on the other hand, provides a basis for judging whether an action should or should not be taken given a certain market value.

Value considerations

The open market value, market value, and most probable selling price are also increasingly used interchangeably to mean value in exchange. The Royal Institution of Chartered Surveyors defines open market value as the best price at which an interest in a property might reasonably be expected to be sold by private treaty at the date of valuation, assuming:

❑ a willing seller;

❑ a reasonable period within which to negotiate the sale, taking into account the nature of the property and the state of the market;

❑ values that will remain static throughout the period;

❑ that the property will be freely exposed to the market;

that no account is to be taken of an additional bid by a special purchaser.

Table 6 C   Value considerations

- ❏ Investment yield
- ❏ Value in use (open market value)
- ❏ Replacement value of the property
- ❏ Land value
- ❏ Performance value of building against other buildings in portfolio and against market

---

Value in use and investment value are generally independent of open market value. Value in use is the worth of a specific property based on its productivity and the particular investor's investment requirements based upon the available investment finance, desired rate of return and other assumptions unique to the investor.

Design and construction sequence

Cost is dictated in two ways: either clients have a maximum sum to invest and want to know how much building work it will buy, or they have a requirement of x sq m of superficial floor area and want to know the construction cost. The owner occupier wants to use the facility for a particular purpose, but the investor and developer have an added dimension because they are also interested in the yield, the value, and ultimately the profit for the project. In general terms, the cost considerations develop in the following way:

Step 1   The investment estimate

Establish the boundary of cost either by the client saying, 'this is what I have to spend', or by asking, 'this is the usable area I require. What is it likely to cost?'

Step 2   The development estimate/development budget

Set the budget limit based upon preliminary forecasts with or without design information. The problem is that the first estimate becomes 'cast in stone' in the client's mind. Ensure there is an adequate contingency allowance to cover unforeseen events.

Step 3   **The design estimate**

Forecast of the construction price for the completed project. Both the likely tender price and the final construction price, and forecasts of future inflation will be incorporated into the estimate.

Step 4   **The cost plan**

Cost plan the project, on a basis of drawn information, dividing the project into elemental categories or trade packages. As the design develops and more information becomes available about the proposed project, so the cost plan can be refined. The cost plan at the early stage might be divided into only 5 or 6 elements, which will be expanded to 20 or 30 elements or trades at a later stage. Increasingly, construction work is sub-contracted and the trade packages relate to the sub-contractors' work.

Step 5   **The cost checks**

Check the design as it proceeds, to ensure it is within the allowable budget.

Step 6   **The tender (the construction estimate)**

The contractor submits the tender.

The listing in **table 6 D** expands the above in more detail.

Obviously there is a difference between a budget, an estimate, and the actual cost incurred. At the early stage of a project, the initial budget is established and it is this first budget figure that becomes indelibly imprinted in the client's mind. The difficulty is that often when the initial budget is being set there is only sketchy information available about the details of the building. The base budget is established upon a cost allowance per unit of area or per person and additional variable costs are added.

The aim is to ensure that the construction price forecast is within the budget cost allowance. The budget is a single price figure, but having recognised that construction work is beset with risk, is it reasonable to expect the construction price forecast to be a single point estimate? A more realistic approach is to adopt the policy that the single point forecast is the most likely price with a range showing the lowest price and the higher price, given that certain conditions exist.

An argument against giving a price range is that the client who is not familiar with the construction industry could abandon the project at the early design stage if the highest price is above the price that he could possibly afford. It is therefore important that the lowest and highest figures have a probability attached to them. It is also worth pointing out that it is much better that the client be aware of likely cost variations at an early stage. Nothing creates a worse impression than for the client to suddenly be presented either with tenders that do not fall within the budget or with a cost overrun during construction.

Table 6 D   The development of cost considerations

Sequence	Action and cost implication
Investment appraisal for financial consideration.	No financial commitment other than consultants' fees. Arrange any project loan finance required. Identify the key risks and appraise the risk exposure for both the lender and the borrower. Establish maximum budget allowance for project.
Development appraisal based upon outline concept design.	Land acquisition and financial commitment to project. Interest charges incurred on any loan finance for the purchase of the land.
Budget allocation based upon the outline design.	Establish maximum budget for construction. Establish the most appropriate method of contract procurement based upon risk, cost and time constraints.
Price forecasts based upon the outline sketch design.	Forecast of the likely construction tender price. Establish revised budget if necessary. Undertake a risk analysis on the budget estimate to consider the most likely and the best and worst eventualities.
Cost plan and cost checks based upon the detailed design.	Develop a cost plan with target costs for the elemental categories. Make any revisions to the price forecasts as the design develops. Check to ensure the elemental categories are within the cost plan allowance. Undertake a risk analysis on the cost plan elements to consider the probability of the budget being achieved. Develop a life cycle cost plan and use life cycle costing to evaluate alternative options.
Life cycle cost planning based upon the detailed design.	Examine the future running costs of the project in use, to ensure they are within the client's brief.
Final price forecast before tender based upon detailed design and production information.	

Table 6 D   The development of cost considerations (contd.)

Tender price.	Contractor's estimate and tender submission. Contractor to identify all the major risks. All the risks will have a response either by avoidance, transfer, reduction or retention. Contract conditions to be scrutinised to ensure the risks have been responded to. Accept the tender and incur full financial commitment for the project.
Management and monitoring of the construction cost.	The contractor is interested to make sure he receives full payment for all the work undertaken. Also to ensure the direct costs are below the tendered price to the client. The contractor is seeking to cover the direct costs plus the overhead allowance and a profit margin. The design team will meet the construction costs during the construction phase. The client will receive a regular update on the anticipated final account. The client's risk exposure will be kept constantly under review.
Final account.	The final commitment for the construction price.
Fitting out and commissioning costs.	Monitoring on behalf of the clients.
Project reconciliation.	Land costs. Construction costs. Professional fees. Fitting out costs. Interest charges on loan finance. Reconcile final cost with budget allowance.

Included in any single point estimate for a proposed project is an allowance for the impact of future inflation. The sums included are a best guess or hunch.Whilst every effort is made to predict inflation, it is outside the control of the design time. The normal approach is to quote prices with an allowance for inflation.

The contractor is interested in his own risk exposure but from a different viewpoint. He wants to ensure that the direct cost incurred on the project is below the tender price, thus ensuring the required profit is achieved. On a lump sum fixed price tender, the contractor has to make assumptions about what allowance to include for inflation, what will happen in the construction environment over the duration of the proposed project and its likely impact upon specialist contractors' prices, and a whole host of other issues. Once the tender has been accepted, the contractor must manage his risk exposure.

Table 6 E   Balancing the budget and the forecast

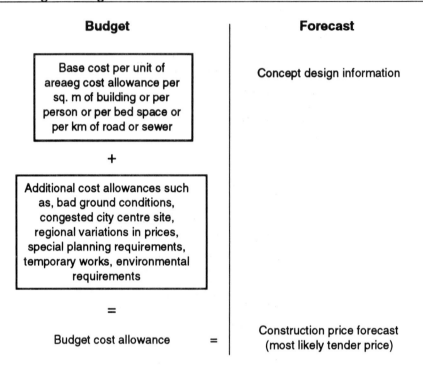

Budget	Forecast
Base cost per unit of areaeg cost allowance per sq. m of building or per person or per bed space or per km of road or sewer	Concept design information
+	
Additional cost allowances such as, bad ground conditions, congested city centre site, regional variations in prices, special planning requirements, temporary works, environmental requirements	
=	
Budget cost allowance =	Construction price forecast (most likely tender price)

TIME DELIVERY SEQUENCE

Figure 6.2 shows the time delivery sequence. Whilst time is money, it must not be assumed that all clients have as a priority the need to obtain the building as quickly as possible. Some public sector clients will be interested in ensuring the building payments are in accordance with their annual financial allocations from Central Government. Other clients may be more interested in certainty of delivery rather than speed.

Indeed, it can be argued that the typical situation places much greater importance on *certainty* of completion either for design or construction. Some events are controllable, but many are not. It is difficult to control the time taken to obtain outline and detailed planning approvals, building regulation or code approvals, and the Fire Officer's approvals. Also, unforeseen events such as exceptionally inclement weather can delay the progress of the work.

Figure 6.2   Time delivery sequence

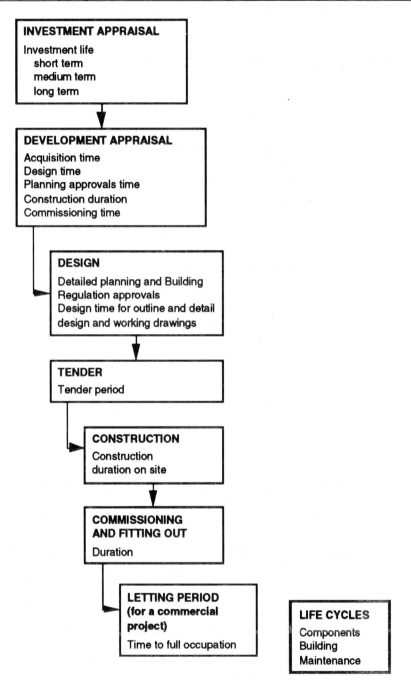

The severe winds that struck the south-east of England in October 1987 were considered to be a once in a hundred years event; the result was havoc on construction sites with consequent delayed completions.

At the start of the project, the client is interested in the start and finish dates. Within the overall design and construction time there are discrete activities, such as the period for design, the period for tendering, and the time spent in construction work on site.

There are sophisticated procedures available, such as critical path scheduling, which can help forecast the time required for various activities. With the advent of computer assisted techniques it is also possible to consider the probability of achieving scheduled deadlines. However, there are very few programs which incorporate a risk management approach to time scheduling.

The delivery sequence for many of the key materials and trade and specialist contractors will be interrupted. The parties on a project often work in a vacuum and fail to see the project from the client's viewpoint. Let us assume the construction of a proposed food store is delayed by one month, owing to unforeseen adverse ground conditions which were not evident from the site borehole reports. The implications are:

❑ specialist groundworks contractor seeks an extension of time and additional money from the contractor;
❑ contractor seeks an extension of time and additional money from the client and all future activities on the critical path will be delayed;
❑ client pays extra for the construction work;
❑ everybody sees time from their own perspective;
❑ client will pay increased professional fees as a percentage of increased construction work;
❑ client will incur additional loan finance interest charges;
❑ one month's trading revenue will be lost for the client.

Obviously a prudent client will have built some float into the overall construction programme, but there is still an associated cost penalty to pay.

To illustrate the complexities and interrelationships between site activities, the time perspective is now examined from the viewpoint of both the contractors and specialist contractors.

Contractors and specialist contractors

The traditional view of the construction process is that once the lowest tender has been accepted and a contract has been signed and agreed between the contractor and the employer, then work commences on site and most of the risk will reside with the contractor. That view might have been true in times when a general contractor employed his own labour to undertake most of the construction operations, but it is not true today.

The recession, the fluctuations in demand for construction, the increasingly onerous conditions placed upon employers when directly employing labour, and the changes in technology which have resulted in new skills and changes in craft skills, have all combined to cause contractors to cease employing large numbers of directly employed craftsmen and labourers. Most of the physical work on site is now carried out by specialist contractors. The increasing use of off-site prefabrication has also brought about the need for new specialist skills for assembly on site. The specialist contractors therefore will have a key role in the future; the best specialists enter into contracts directly with clients.

The roles played today by specialist contractors have evolved by default rather than by design, resulting in their accepting responsibility and risk which they are not necessarily well equipped to handle. Many firms sign onerous contract conditions because if they refuse, they do not get the work.

Competition is based upon price and, apart from some of the major companies, there is very little customer loyalty. Some long term relationships are built up between contractors and specialists, but in Western culture it is the price that eventually determines who does the work.

Over the years, the extent of specialisation has increased. There are an increasing number of specialist firms with product skills, that is, skills relating to a product or series of products. While craft skills are still the most important, the mix of skills is changing, on repair and maintenance work and smaller projects there is a need for multi-skilling.

The specialist contractor can fulfil a number of roles. The specialist is increasingly being called upon to provide an input into design. Some specialists, such as the lift manufacturer, will provide a complete service by designing the installation, manufacturing some of the components, being responsible for assembly and fabrication of the lift system, and installation into the building, as well as providing an ongoing maintenance service. Similarly, the curtain wall specialist will provide a detailed design to meet the architect's design concept and specification. At the other end of the spectrum, the bricklaying specialist contractor will provide the labour only to lay the bricks on site.

The permutations across the spectrum are wide ranging. **Figure 6.3** provides a risk check list. It shows the process divided into the four categories of design, materials manufacture, assembly, and installation.

The risks, divided into the money, time and technical risk categories are then examined within each category. For instance, at the installation phase the specialist contractor must ensure that he performs to the construction programme, that the work is undertaken in the budgeted number of visits to site, and that no delay is caused to other specialist contractors.

A major difficulty faced by the specialist is that the activities are rarely independent and mutually exclusive. For instance, if the electrical specialist contractor is delayed because of the non availability of key electrical equipment being imported from Japan, and the activity is on the critical path, then the project will be delayed. All the specialist contractors affected by the activity will have to reschedule their labour and material deliveries.

Figure 6.3   Risk and the specialist contractor

Process	Money	Time	Technical
Design the product, Desgn for the project	• Submit estimate for undertaking the work • Cost/price balance for the design element	• Meeting the design programme	• Design liability • Fitness for purpose • Meeting the stipulated technical standards
Materials manufacture, supply	• Cost/price balance for the manufacture and supply element • Contract risk	• Meeting the stipulated delivery schedule for the site or assembly/ fabrication plant	• Ensuring the correct quality is maintained • Product liability
Assemble, fabricate	• Cost/price balance for the assembly element • Contract risk • Estimating errors • Inflation	• Meeting the stipulated delivery schedule for the site	• Availability of materials • Organisation and management of sub-contractors
Install, licence to install only, labour only	• Cost/price balance • Estimating errors • Inflation • Cash flow • Possibility of incurring liquidated damages for late delivery of work • Getting paid the right amount in the final account at the right time • Contract risk • Latent defects price risk	• Meeting the overall construction programme • Meeting the required number of visits to site • Minimising disruption due to overlap of work with other trades • Delay by exceptionally inclement weather • Possibility of causing delay to other specialist contractors	• Quality of workmanship • Organisation and management of the project team • Labour relations • Latent defects upon completion caused by poor workmanship or technology • Safety of personnel

In order to illustrate how complex these activities can be, **figure 6.4** shows diagrammatically the sequence for two different types of specialist contractor, the curtain wall specialist contractor and the painting and decorating contractor.

Figure 6.4 Curtain wall specialist contractors' approvals

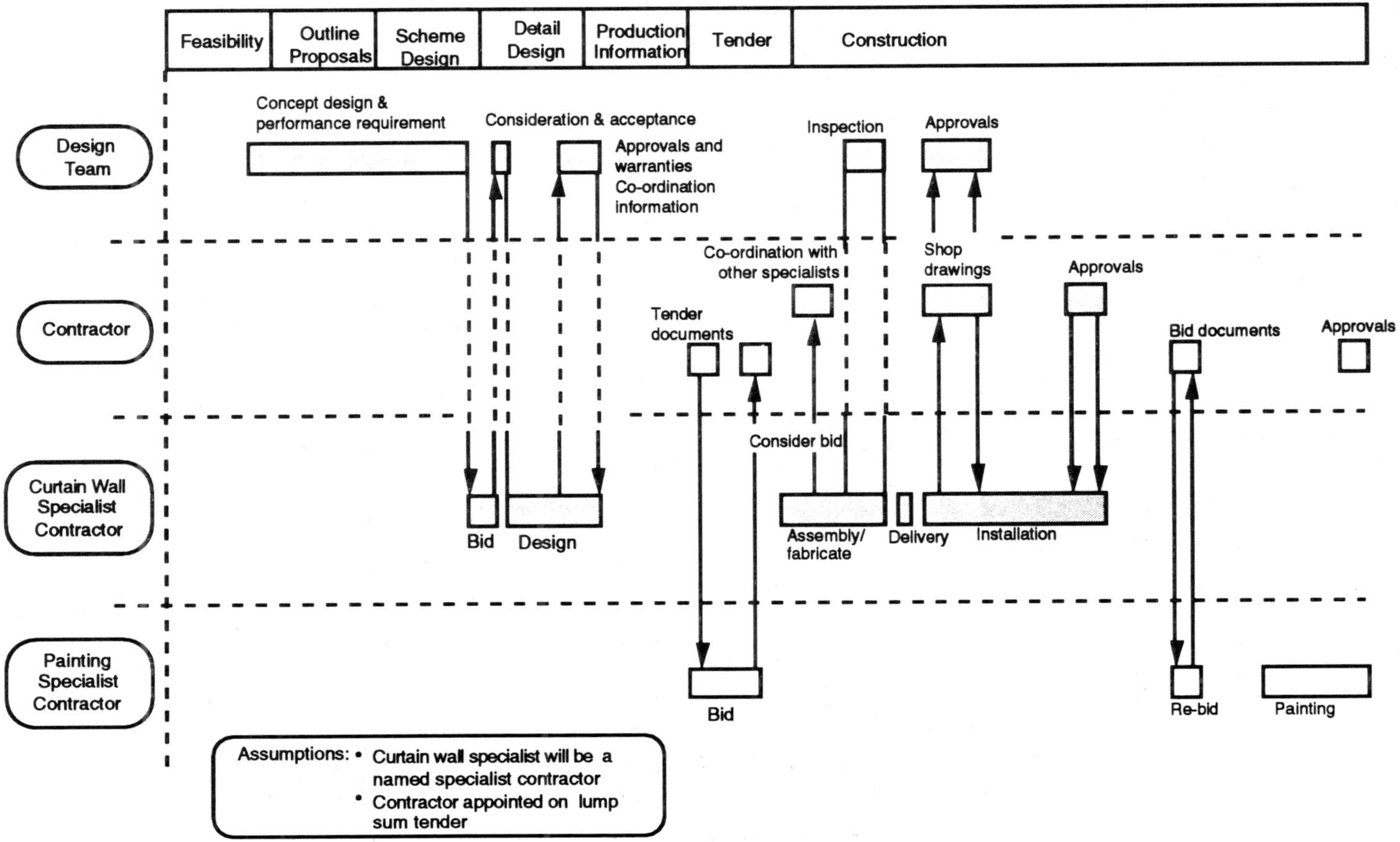

In the example, it has been assumed that the curtain wall specialist will be a named specialist contractor and is responsible for the detailed design, manufacture, and installation, whereas the painting contractor is a sub-contractor to the general contractor.

There are many stages where slippage on the design and construction programme can occur. The specialists shown are only two of a large number of specialist contractors. Where design manufacture and installation are involved, the sequence of events is even more tortuous. However, by careful planning and good fortune, projects are completed on time.

So, how can the risk of a time overrun be minimised? The first rule is good planning. The second rule is the identification of the high risk activities and the equitable allocation of these risks. Critical path scheduling identifies the activities on the critical path; it also allocates the free float available. The allocation of a probability value to the critical path activities will show the probability of completion within the scheduled time.

There is a potential pitfall with this approach. In theory, critical path analysis is an indispensable tool for construction planning. In practice, it is used at the strategic level of planning but rarely used on site because of its 'apparent complexity' and the need to keep the programme up to date. Invariably the site staff will use bar charts taken from the critical path analysis. Risk analysis techniques have the same potential for producing a gap between a good theoretical approach and a lack of application at the project level. There is a difference, however, because the risk analysis for time planning is appropriate primarily at the strategic level.

TECHNICAL DELIVERY SEQUENCE

Throughout the whole technical delivery sequence quality is of prime importance, whether it be the quality of the investment or the quality of the end product.

The range of interests in the technical delivery sequence is very wide, as illustrated in **figure 6.5**. There is no magic formula to manage the technical risks, wherever possible, the risks should be quantified and analysed. Most importantly a risk table needs to be prepared showing the source, effect, and method of risk allocation.

Some aspects, such as safety on site, are of paramount importance. The key to safety is proper training and commitment to a safe work place. Everybody is responsible for safety but too frequently it is seen as being somebody else's responsibility. The safety officer's task is to identify risky activities and to devise a safe method of working.

To show some of the risks to be considered by a professional person in his or her everyday activities, a case study of the building surveyor's perspective on risk is shown below.

Figure 6.5   Technical delivery sequence

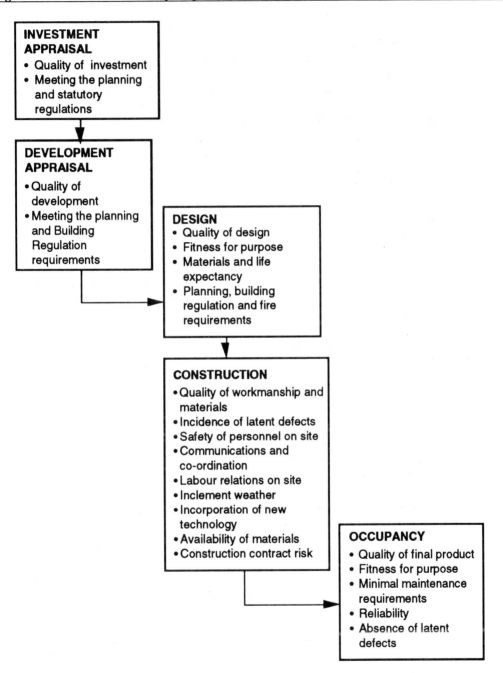

A case study of the technical risks faced by the building surveyor

The role of the building surveyor encompasses parts of other construction disciplines. Some building surveyors, for example, will operate in highly specialised areas such as timber use and preservation while others will have broader based skills and work in the areas of rehabilitation, structural surveys, condition surveys and dilapidations.

As the areas of operation become more diverse, so the building surveyor will be required to make decisions and offer advice on topics for which he has only general or passing knowledge; this ultimately leads to high risk exposure for the building surveyor.

Generally, the building surveyor will be involved with new or existing buildings of any age and will undertake surveys for either known or unknown users.

When making assessments of material performance, the surveyor will have varying amounts of information available on which to base an opinion. Involvement at this early stage involves risk, as the building surveyor will have to base his assessments on some information which cannot necessarily be confirmed as illustrated by **figure 6.6**. When undertaking a structural survey of an existing building, the building surveyor deals with confirmable and non-confirmable data. These are listed to show what assumptions have to be made, particularly where the workmanship is hidden. An example would be a five year old timber framed detached house where the condition of the external framing can only be determined by removing a substantial portion of the exterior cladding. The vendor of the house would not readily agree to part of the cladding being removed, yet the purchaser still wants an opinion as to the likely condition of the timber.

Given the correct specification and supervision of site works, the risk of incorrect assessments of performance can be considerably reduced. However, without constant checking, there will always be the risk of failure due to poor workmanship or incorrect use of materials.

An important role for the building surveyor is in undertaking instructions relating to advice on the standard of repair or structural suitability of existing buildings. This work usually involves giving opinions and advice for two broad areas:

❑ the actual performance of the building structure, materials and workmanship;
❑ the future performance of the above, together with predicted life cycles and replacement/repair costs.

In the main, the surveyor will be required to make assumptions and state opinions on parts of buildings where the information available is either adequate, unobtainable or contradictory.

Figure 6.6   Confirmable and non-confirmable data for the building surveyor

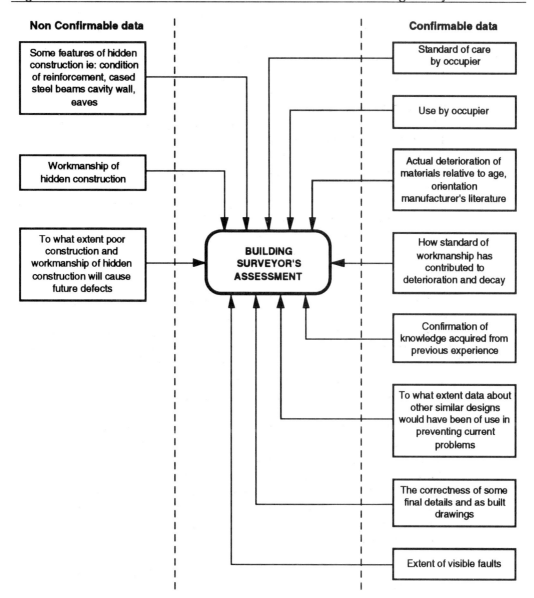

Despite the burdens placed on surveyors by recent legal decisions, it
must be remembered that the diagnosis of building performance is not a
precise scientific process and can be carried out with only a degree of
experience and expert judgement. Hence there must always be an element
of doubt and risk in diagnosis and advice to clients. The aim must be to
minimise that risk exposure.

Figure 6.7 shows the identification of a problem, dampness at the ground floor level near skirting boards. The building surveyor divides the possible causes into broad categories and then seeks to identify answers to a number of questions within these categories. Some of the questions are straightforward and have very little risk attached to them. However, consider the problem of rising damp caused by the cavity having been bridged at the damp-course level. Whilst an endoscope can be used to inspect the cavity, it is difficult without demolishing part of the wall to be specific about the standard of workmanship inside the cavity for the whole building. The building surveyor will react to the dampness caused in that part of building where there are signs of dampness. He cannot be sure dampness will not occur in another part of the building at a later stage.

Figure 6.7   Seeking the cause of the problem

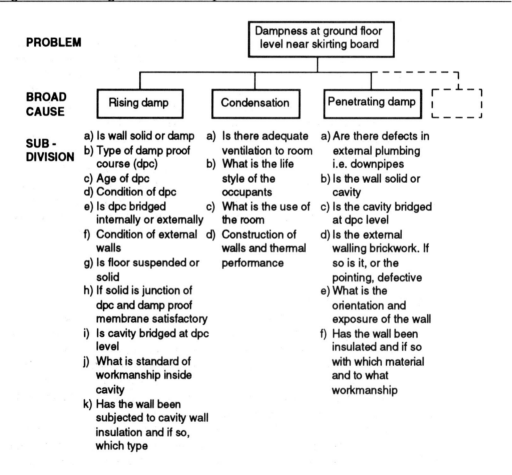

7

SENSITIVITY ANALYSIS, BREAKEVEN ANALYSIS, AND SCENARIO ANALYSIS

SENSITIVITY ANALYSIS

Sensitivity analysis is a deterministic modelling technique which is used to test the impact of a change in the value of an independent variable on the dependent variable. It does not aim to quantify risk but rather to identify factors that are risk sensitive. Sensitivity analysis provides answers to a whole range of 'what if' questions. For example, what happens to the construction price if the allowance for future inflation is underestimated by 1%, or 2%, or 3%? What happens to the construction price if the work on site is reduced by 3, 4 or 5 months?

Sensitivity analysis enables us to test which components of the project have the greatest impact upon the results, thus narrowing down the main variables to be considered. The technique is widely used because of its simplicity and ability to focus on particular estimates. It does not however actually evaluate risk, the decision maker must still assess the probability of an event occurring.

BREAKEVEN ANALYSIS

This technique is an application of sensitivity analysis. It can be used to measure the key variables which show a project to be either attractive or unattractive. A simple example for a project would be examining the critical rate of return with the cash inflow and initial cash outflow, the capital cost, the rate of inflation, the discount rate, and with a rent review every three years with the new rent being based upon the annual rate of inflation in the year preceding the review plus 2%.

The rate of return is calculated by finding the appropriate rate which equates all future cash flows with the initial capital cost. The net present value criterion merely states that a project is worth undertaking if the present value of all future discounted cash flows is greater than, or equal to, the initial capital cost.

Table 7A shows the data for a proposed investment with various assumptions. The results show a net present value of -£2,555,848 which means that the project is not a good investment. The rental income would need to be £770,000 per annum in order for the project to be worthwhile if the other values remain constant.

Table 7 A   Investment appraisal of an office building

Capital cost (land construction, professional fees, taxes)	£6 million
Cash inflow (rental income)	£700,000
Cash outflow costs (running costs of building)	£200,000
Assumed annual inflation rate	4% pa
Rent review period	3 years
Rent review allowance above inflation in the final year preceding the review	2% pa
Discount rate	12.5%
Time horizon	30 years
Net lettable floor area (initial rent £10 per sq. ft)	70,000 sq. ft

This analysis can be extended to test, one at a time, the other crucial variables to the break even rate. For example the inflation rate, the discount rate, and the running costs could be varied to check the impact upon the break even rate of return.

SCENARIO ANALYSIS

This is a rather grand name for another derivative of the sensitivity analysis technique which tests alternative scenarios; the aim is to consider various scenarios as options.

When undertaking a scenario analysis the key variables are identified together with their values. For instance, at the early concept stage of a project the architect might be looking at various different layouts of the building to optimise the net to gross floor area by minimising the amount of circulation and balance area, i.e. the area taken up by toilets, corridors, lift wells, and ducts. Similarly, the provision for car parking might be under consideration. The quantity surveyor could present various scenarios showing the impact on the cost change for the change in floor area, forecasts of inflation, and the expenditure on roads and car parking that might be stipulated by the local authority.

Figure 7.1   Scenario analysis for proposed office building development appraisal stage

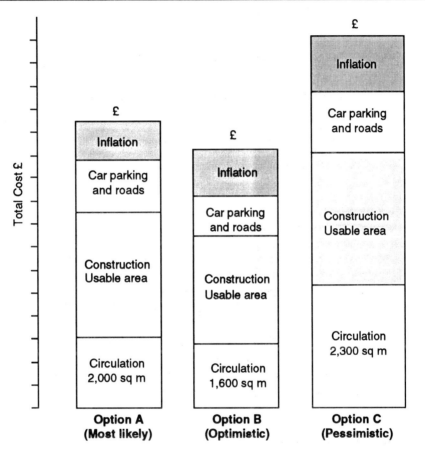

The scenarios should be based upon the most likely, the optimistic, and the pessimistic forecasts. **Table 7B** shows the alternative scenarios. The gross superficial floor area will vary with the circulation space requirement. The extent of the modification to the roads will depend upon the local authority's stipulations which, at the early stages of the design, may not be known in detail. Because the specification is still being refined the range of the anticipated construction cost is from £950 to £1,100 per sq m.

Table 7 B   Scenario analysis for proposed office building

	Option A Most likely	Option B Optimistic	Option C Pessimistic
Circulation space area required to meet the client's requirements	1,000 sq. m	1,600 sq. m	2,300 sq. m
Net usable floor area of the building stipulated in the brief	7,000 sq. m	6,600 sq. m	7,300 sq. m
Gross superficial floor area of the building	5,000 sq. m	5,000 sq. m	5,000 sq. m
Construction prices forecast for building cost per sq. m at the fourth quarter 1988	£1,000	£950	£1,100
Inflation allowance for a 12 month design time and 12 month construction period	5% pa	4% pa	8% pa
Cost of providing car parking and modifying the existing roads to meet L.A. requirements	£500,000	£400,000	£800,000

The results represent the range of possible outcomes. At this stage no quantitative measure of the possibility of each outcome has been placed upon the option, although clearly the client could ask for such a measure.

SENSITIVITY ANALYSIS - AN APPLICATION TO LIFE CYCLE COSTING

An example of the use of sensitivity analysis is in the consideration of life cycle costs for a project. Life cycle costing, by definition, deals with the future yet the future is unknown. Recurrent costs such as maintenance, replacement and cleaning costs are only estimates no matter how precise and reliable the data on which they are based. Similarly, other essential components of life cycle costing, such as the rate of exchange (i.e. the discount rate) between future costs and their present values, replacement cycles of individual components, and the life cycle of the building itself cannot be assessed with certainty.

It is probably fair to claim that one reason, or, more properly, excuse, for the relatively slow introduction of life cycle costing methods to the building industry has been a feeling that life cycle cost estimates are in some sense inaccurate, or based merely on guess-work. If this view is to be challenged effectively, simple, practical techniques must be developed that address risk and uncertainty explicitly, and give the decision-maker comprehensive information on which to base their judgements.

A major objective of life cycle costing is to rank competing projects, whether these be competing designs, or competing finishes to a floor space. Sensitivity analysis of the life cycle cost may indicate that the ranking of the options being considered is unaffected by variation in a particular parameter. There is little value in trying to improve the estimate of that parameter.

As was indicated at the start of this chapter, sensitivity analysis is generally used to identify the impact of a change in a single risky or uncertain parameter used in the calculation of life cycle cost (LCC) such as, for example, discount rate, initial capital cost, or running costs. It identifies the sensitivity of LCC to variation in each of these parameters and this has two major uses.

Firstly, it indicates for a particular option, the certainty that can be resided in the LCC calculation based on best estimates of all parameters. If the decision-maker is interested in reducing uncertainty or risk exposure, then sensitivity analysis will identify those areas on which effort should be concentrated in order to improve the parameter estimates.

Secondly, sensitivity analysis indicates in the comparison of alternatives the conditions under which the ranking of these alternatives will change. Assume that the decision-maker's primary objective is to get the alternatives in the correct order. Then it may not matter that LCC, for these alternatives, is sensitive to a particular parameter if the ranking of the alternatives does not change when the parameter is varied through its expected range - one such example is illustrated later in the case study. In this case, effort should be concentrated on improving information with respect to those uncertain parameters that do give rise to a change in ranking.

There are several ways in which the results of a sensitivity analysis can be presented, the simplest of which is to compute a sensitivity table. However, if several variables are changed, a graphical representation of the results is most useful as it quickly indicates the most sensitive or critical variables. In addition, a graphical representation is useful in identifying relative uncertainty when faced with the choice between competing options and in identifying areas of further risk and uncertainty between those options.

A particularly effective graphical presentation of sensitivity analysis - the spider diagram - has been suggested by Perry and Hayes (1985) and has already been discussed briefly in Chapter 4. This has been developed primarily in the context of civil engineering, but has obvious applicability to the building industry. It is best illustrated by means of a hypothetical project as in **figure 7.2**. The method of construction of a spider diagram has been illustrated in Chapter 4 and is repeated here for ease of reference:

i)   Calculate life cycle cost using best estimates of all parameters.

ii)   Identify the parameters subject to risk and uncertainty.

iii) Select one of the risky parameters and recalculate LCC assuming that this parameter is varied by £ x% where x lies in some pre-defined range. This should be carried out in steps within this range, e.g. recalculate LCC assuming that the discount rate is changed by +1%, +2%, ...+5%, and -1%, -2%, -... -5%.

iv)   Plot the resulting LCCs on the spider diagram, interpolating between each value. This generates the line labelled 'parameter 1' in **figure 7.2**.

v)   Repeat stages iii) and iv) for the remaining parameters that have been identified as risky.

Figure 7.2   Sensitivity analysis spider diagram

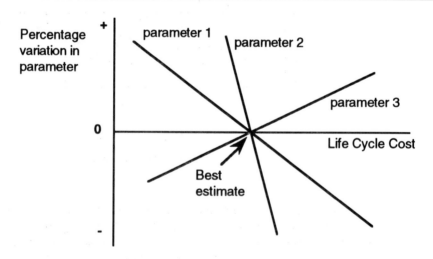

Each line in the spider diagram indicates the impact on LCC of a defined proportionate variation in a single parameter that has been identified as having some risk associated with its estimate. The flatter the line the more sensitive LCC will be to variation in that parameter. For instance, it can be seen from **figure 7.2** that a variation in the estimate for parameter 3 would have a much greater impact on LCC than an identical variation in parameter 2.

In relation to this last point, figure 7.2 does not give any indication of the likely range of variation of each of the risky parameters. This is overcome by incorporating probability contours into the spider diagram. These probability contours are constructed by subjectively identifying the range within which a particular parameter is expected to lie at each level of probability. For example, it might be estimated that there is a 70% probability that the discount rate will lie in the range between +8% and -6% of the best estimate, and a 90% probability that the range is between + 10% and -8%. **Figure 7.3** illustrates a spider diagram with 70% and 90% probability contours added.

Figure 7.3   Probability contours

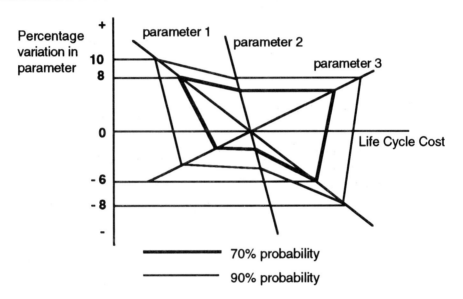

Criticism can be made of the use of probability contours precisely because they are subjective estimates of the likely range of variation of the relevant parameters. We are not convinced that such criticism has any strong validity. Sensitivity analysis is a management and decision-making tool. Our feeling is that the judgements underlying the placing of the probability contours is precisely the kind of judgement that management can be expected to make.

In addition, the placing of probability contours can be made rather more accurate if additional statistical information is available on the risky and uncertain parameters. In particular, assume that the 'best estimate' of a risky parameter is its mean value, and that an estimate can be made of that parameter's standard deviation. Then if the underlying probability distribution of the parameter is assumed to be the normal distribution, one can take advantage of the statistical properties of the normal distribution. Specifically, it is known that 68% of the possible values of a normally distributed parameter lie within plus and minus one standard deviation of the mean, 95% between plus and minus two standard deviations, and 99% between plus and minus three standard deviations. For moderately skewed distributions these values hold approximately, while for a log-normal distribution (which exhibits the kind of skewness we might expect to find in cost estimates) these values hold in logarithms of the parameter and can easily be transformed into standard values.

Using these properties, it should be clear that probability contours can be placed with some confidence - always provided, of course, that estimates are available of mean and standard deviation.

The discussion thus far has considered sensitivity analysis as it applies to a single option. A prime function of LCC, however, is to compare and rank alternative design solutions - whether this be for a complete building or for a particular component such as the choice of finishes. Sensitivity analysis, using spider diagrams, provides a particularly useful method for guiding such comparisons and rankings.

This is best illustrated by means of the hypothetical example of **figure 7.4** which might refer to choice of floor finish to be included in a particular functional space. Best estimates of the relevant parameters lead to Finish A being preferred, having the lower life cycle costs as compared with Finish B - points A and B in figure 7.4. But Finish A is much more sensitive to variation in the uncertain parameters; as can be seen from the degree of sensitivity within the 70% probability contour. The reason for the change in the hatching on the diagram is because parameters 1 and 2 in the option A and B cross, thus showing the greater sensitivity of option A parameters. If a primary objective of the decision-maker is to 'avoid surprises' then there would be considerable justification in preferring Finish B since it offers much greater cost certainty.

Figure 7.5 illustrates a more detailed analysis of sensitivity analysis applied to the rank ordering of the options under consideration. Again, we illustrate an example in which option A has the lower life cycle costs on best estimates. As can be seen, if parameter 1, which in this case represents discount rate, were to vary by anything in excess of 5% above its best estimate, the rank ordering of the two options would change. Similarly, if parameter 3 were to vary by more than 7% of the best estimate value then again ranking of the two floor finishes changes.

This type of comparison can be repeated for each parameter to indicate those parameters for which the rank ordering of the two options would change. This is indicated by the shaded areas in **figure 7.5**.

Figure 7.4   Comparison of options 1

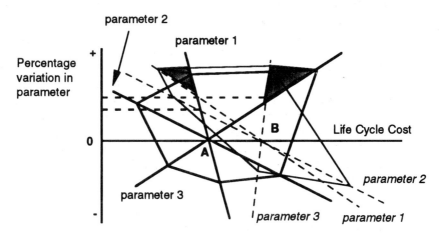

The greater the extent to which the rank ordering of the options under consideration would be changed by a parameter variation within the chosen probability contour, the less clear it becomes to reject option B in favour of option A.

Figure 7.5   Comparison of options ll

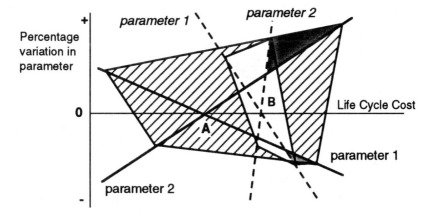

It is apparent from our discussion, and in particular the examples given in **figures 7.4** and **7.5**, that sensitivity analysis does not provide a definitive method of making a choice between the competing options. However, even as currently developed it is an essential component of managerial decision-making. In particular, the use of spider diagrams, together with rank ordering of the options by applying probability contours and analysing the amount of variability necessary in individual parameters to change that rank ordering, provides the decision-maker with a useful decision tool.

One limitation of sensitivity analysis is that, within its current applications, it is univariate - only one parameter is varied at a time. It is possible to overcome this to at least some extent by borrowing contour analysis techniques used, for example, in economics and topography. Assume that two parameters have been identified from a 'standard' sensitivity analysis as being particularly important in determining the LCC of a particular option. Further assume that an increase in parameter 1 increases LCC and a decrease in parameter 2 also increases LCC (parameter 1 might be project life and parameter 2 the discount rate). Suppose that best estimates of these two parameters generate LCC of £0.5m. It will be possible to find other combinations of project life and discount rate that will also generate an LCC of £0.5m: by increasing project life and increasing the discount rate, or decreasing project life and decreasing the discount rate.

RISK ANALYSIS USING MONTE CARLO SIMULATION

PROBABILITY ANALYSIS - EXTENDING THE SENSITIVITY TECHNIQUE

We now turn to an approach to risk and uncertainty that takes explicit account of the fact that all risky and uncertain parameters can be expected to vary simultaneously.

Probability analysis is a powerful tool in investigating problems which do not have a single value solution. Stochastic simulation in the form of a Monte Carlo simulation is perhaps the most easily used form of probability analysis. It makes the assumption that parameters subject to risk and uncertainty can be described by probability distributions. The Monte Carlo technique makes use of these probability distributions to generate a number of simulations of the desired overall cost estimate.

HOW IT WORKS

The simulation of a project uses uncertainty in the estimating process. Firstly, a project is broken down into activities or packages with an assessment of reasonably foreseeable 'optimistic' (or lowest) and 'pessimistic' (or highest) values or times along with the 'most likely'. The most likely value becomes the peak value of the distribution. Seldom is it possible by the use of historical data or expert judgement to define more than these three points, nor is it even worth attempting. Estimating is a case where practicality of approach and sensitivity of results militate against analytical overkill. Most of us cope with the uncertainty in practice by describing it with words like 'between x and y'. When we have to be more definitive we will include contingency sums to cover the unknowns we are dealing with. Uncertain things are fuzzy values, not precise values.

A simple example can be used to show the difference between conventional single value analysis and when simulation is used. In deterministic analysis, the single value estimate produces a single value overall result. Whereas, in simulation, one pass through the project occurs with independent samplings from the distributions defined for the activities. Some may show values towards the optimistic end of their ranges, while others may experience pessimistic values. After the first pass through the project, the computer performs the entire operation again and again. The computer uses a random number generator to choose the number within the range of values. It uses the distribution to decide the frequency from which the number is selected.

As in real life, there is virtually no case where all activities experience their best or worst values during a project. The overall result is a distribution of reasonable foreseeable project costs which give an indication of the range of prices likely to be encountered. The simulation is using the three point values and describing everything that goes on between the values. In effect, the probability distribution weights the three values that have been provided.

USING MONTE CARLO SIMULATION IN THE COST PLANNING OF A BUILDING

The cost of construction work is a combination of what the client is prepared to pay and the price at which the contractor is prepared to undertake the work in order to show an acceptable level of profit. Prices can be considered as residing in a family of prices where there will be both extremes and a most likely price.

Clients generally understand that the single value price forecast made at the design stage of a proposed building represents, under normal conditions, the most likely tender price. However, they lack a means of gauging the possibility that the tender price will either exceed or be less than the price forecast.

It is axiomatic that the price forecast is merely an estimate of the tender price. Nevertheless, clients must have confidence that the price forecast is neither too optimistic nor too pessimistic an indication of the most likely price. If estimates are too optimistic, clients will waste time and resources on development plans that will be impracticable when tenders are received. On the other hand, too pessimistic a view will have the effect of depressing investment in construction and, in consequence, investment in other economic activities.

The aim of this chapter is to describe a technique which identifies the probability distribution from which a price prediction has been taken. It can initially be used by the client and consultants in assessing a particular price prediction, and, eventually, in deciding upon the feasibility of the construction project to which the prediction refers. The general methodology is applicable at any stage of the investment appraisal, development appraisal, or design process of a project. We have, however, illustrated its use by applying it to the cost plan prepared at the scheme design stage.

The single factor that characterises all price forecasting is uncertainty. Price prediction is not a precise scientific exercise, but an art which involves both intuition and expert judgement. Despite the undoubted desirability of an unbiased price prediction, there exists no objective test of the probability that a particular forecast will be achieved. Since a price prediction is the sum of many parts, any such objective evaluation of its precision is possible only by the use of statistical techniques.

Probability theory allows future uncertainty to be expressed by a number, so that the uncertainty of different events may be directly compared. Information about the probability of a future event occurring, or a condition existing, is generally presented in the form of a probability density function. If, then, we can obtain some indication of the probability density function to which a particular price prediction belongs, we have available a test of the likelihood that the estimate is unbiased.

ESTIMATING AND PRICE PREDICTION: AN OVERVIEW OF CURRENT PRACTICE

When we estimate construction prices for a proposed project, we often look at past projects as the data base that is adjusted for future projects. Construction prices used in forecasting are often based on the analysis of a small sample of historical projects for which there are cost analyses or some form of cost breakdown, and which bear a close resemblance to the proposed building project. It is assumed that the tender price of an item to be built in the future can be determined by analysis and adjustment of the tender prices of analogous items built in the past.

In some situations there are no past cost data available, in which case experience and skill will play a role in collecting information to estimate a price. Several factors interact and affect the reliability of the price forecast, including:

- ❑ the extent of design information available (ambiguity in both the design and the prediction go hand in hand, whatever the method of prediction);
- ❑ the availability of historical price data related to the type of project under consideration;
- ❑ the familiarity with the type of project in hand and projects of a similar nature.

We shall not be discussing these points in detail, but it is important to note that the prices used in the forecast can only be as good as the sample on which they are based. Further, the 'goodness' or otherwise of the sample is at least a two-dimensional concept. Other things being equal, it is always desirable that the sample should be as large as possible. On the other hand, it is important that the sample contains only construction projects that resemble the proposed project. In other words, the sample should be reasonably homogeneous with respect to the major cost significant features of the project.

It has been suggested that a significant improvement in reliability can be obtained if historical price data are drawn from several buildings rather than from one, even if this means sacrificing some comparability. The exception to this is where there is a single identical completed building. Since the available database is finite, for instance in-house cost analyses of completed buildings, or, in the UK, the Building Cost Information Service (BCIS) cost analyses, a trade-off is imposed between sample size and homogeneity. Unfortunately, the precise nature of this trade-off is unknown, but as the uncertainties are high, a limited number of samples (for instance, fewer than five) are probably inadequate.

A word of caution is worth noting here. Often, cost data relate to the analysis of the tender price, whereas the information needed is the final account price of the completed project. Furthermore, cost data mask the impact of regional differences in construction prices and differences in the size, quality, complexity, and buildability of projects. Professional skill and judgement is therefore needed in the careful selection of projects similar to the proposed project.

COST PLANNING AND RISK ANALYSIS

Interdependence of items

A further point to be considered is the interdependence of the elemental categories used in cost planning. Research has shown that certain of the elemental categories are not mutually exclusive. Some of the elements will be interdependent, for instance the cost of the electrical installation is likely to be higher in a building with air conditioning and where the mechanical services cost element is high. Any risk analysis programme does not take account of this interdependence, other than by examining the correlation coefficients. The only practical way this can be accommodated is by careful examination of the data. Inevitable, this is a criticism of any technique which uses historical data for cost planning. There are numerous different estimating techniques used at the design stage. We shall concentrate on one of these techniques, elemental cost planning using the Building Cost Information Service (BCIS) elemental categories. **Figure 8.1** shows, in diagrammatic form, an outline of the cost planning process.The elemental unit quantities are measured for each selected category and the historical sample is used to generate unit price rates, which are based upon the arithmetic mean or some measure of central tendency. The derived unit price rates for each element are multiplied by the appropriate elemental quantity to build up the final tender price prediction of the likely tender price.

Figure 8.1   Outline of the cost planning process

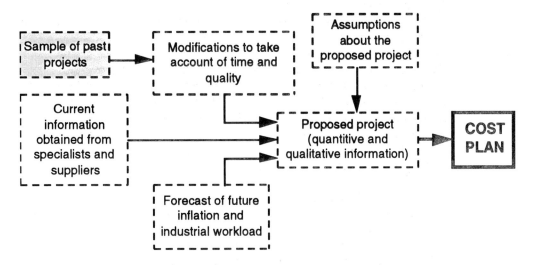

In this prediction procedure, each calculated mean unit price rate represents a sample mean from a probability distribution. A different set of historical cost analyses or price data is likely to generate a different mean unit price rate, that is, a different sample mean from the same probability distribution. It is to be expected, of course, that there would be some offsetting changes in the mean unit price rates - some rising and others falling - but the resulting total price forecast represents only one possible value from a family of values.

No direct evidence can be obtained, at any time, of the way in which a prediction would be affected if it were based on a different, but equally homogeneous, sample; if such samples were available, there would be nothing to be gained by excluding them from the initial forecasting process. Indirect evidence of the probable variability of a prediction can be obtained, however, by using the variability of the calculated mean unit price rates. No matter how homogeneous the sample from which the elemental mean unit price rates are calculated, the individual unit price rates will exhibit some residual variation about the 'true' mean unit price rates as long as the sample contains more than one building. It should be possible to use these residual variations to generate a picture of the probability distribution of the final prediction.

In other words, since each estimated unit price rate is drawn from a probability distribution, it follows that the overall forecast is also a member of a probability distribution, the characteristics of this distribution being determined by the characteristics of the individual distributions for each elemental category. Using the analysis outlined below, it is possible to approximate the probability distribution of the overall prediction to identify the characteristics of the family from which it is drawn. This in turn will allow us to identify:

❑ the probability that the contractor's tender price will not exceed the prediction;

❑ the most likely range within which the contractor's tender price will lie.

RISK ANALYSIS USING PROBABILITIES

Some risk cannot be confidently costed at the design stage, such as the effect on cost that exceptionally inclement weather will have on a project due to the foundation work commencing on site in December. Whereas the majority of risks arise from matters where there is a lack of information, for example, insufficient design and specification information at the early stages of design. As more information becomes available during the design phase, so many of the risks can be resolved until, just prior to a tender being sought, the estimate of construction cost contains only residual levels of risk.

A straightforward approach to including an allowance for the risk is to identify a list of risk items and assign each item with the probability of the event occurring and to give a three point estimate of:

❑ a most likely price;

❑ the lowest price;

❑ the highest price.

A typical risk item might be the probability of the need for a new gas main to be laid on site. At the design stage the existing gas main had not been exposed hence the condition was not known. The most likely price might include an allowance for some modification to the existing main, whereas the lowest price could assume no work to the existing main is required, and the worst case, that substantial work is needed to modify the main. Probabilities for each event would be assigned to each event such that there is a 0.50 chance that some modifications are required, a 0.30 chance that no work is required and a 0.20 chance that substantial work will be needed.

The assessment of probability can rarely be an exact science, therefore expert judgement and intuition are required. In the above situation should the client be made aware of the worst likely eventuality or should an 'average risk allowance' be included in the budget forecast? For example:

Item		Price	Probability	Allowance
A	Some modifications to the existing gas main	£5,000	0.50	£2,500
B	No modifications required other than inspection	£2,000	0.30	£600
C	Substantial modifications required to the gas main	£15,000	0.20	£3,000
TOTAL				£6,100

Consideration must be given to what premium should be added to the budget price to ensure that the client has some comfort that all the eventualities have been taken into account. It might seem illogical to include £6,100 which would cover both options A and B, but it must also be borne in mind that even the prices of £2,000, £5,000 and £15,000 will be informed best guesses based upon the information available. There is a 0.20 probability that option C will be required. A more sophisticated approach uses a process called Monte Carlo Simulation.

RISK ANALYSIS USING MONTE CARLO SIMULATION

We have used elemental cost planning as the basis for illustrating risk analysis. It is equally applicable to other forms of prediction and estimating for building work where a sample of historical prices is being used. For example, forecasting the construction duration using activities and time using simulation is frequently undertaken.

Risk analysis generates hypothetical mean unit price rates for each elemental category in the cost plan for the proposed building. These hypothetical rates are taken from probability distributions with the same statistical properties, that is, probability density functions, as those which characterise the original sample data from which the mean unit price rates were estimated. The hypothetical rates are then used to build up a total price forecast for the proposed building. If this exercise is repeated a sufficiently large number of times, it will be possible to obtain a picture of the probability density function which characterises the total price, and so to identify the most likely total price.

CONSIDERING SOME PROBABILITY DISTRIBUTIONS

The process of selecting a probability distribution sometimes presents difficulties for the practitioner. To choose the correct probability distribution follow three rules:

1   List everything you know about the variable and the conditions surrounding the variable.
2   Understand the basic types of probability distribution.
3   Select the distribution that characterises the variable under consideration.

Remember the basic rules for the measure of central tendency:

❑   the mean, whose greatest weakness are the extreme values which are sometimes unrepresentative of the rest of the data. With the mean, the calculation gives the same 'weight' to each data item.
❑   the median, which is the mid-point in the distribution - half lie above the median and half lie below.
❑   the mode, which is the most frequently occurring value. Whatever the shape of the distribution, the mode is always located at the highest point, as **figure 8.2** shows.
❑   the standard deviation is the square root of the average squared deviation from the mean.

Figure 8.2   A probability distribution

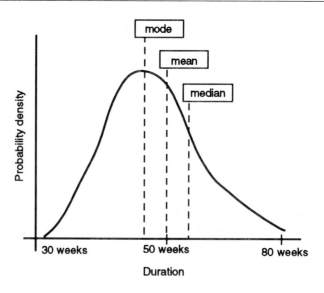

Distributions can either be *continuous* or *discrete*. Most construction industry activities are continuous where values can arise within a given range. Discrete distributions involve values where only specific values in a range can arise.

Common distribution types

- ❑ Uniform
- ❑ Triangular
- ❑ Normal
- ❑ Poisson
- ❑ Binomial
- ❑ Lognormal
- ❑ Exponential
- ❑ Geometric
- ❑ Hypergeometric
- ❑ Weibull
- ❑ Beta

It is beyond the scope of this book to give a description of all these distributions, most good statistics books will give a comprehensive description. In most instances, choosing the best distribution for an analysis is very straightforward. Four of the commonest distributions are discussed below.

Uniform distribution

In the uniform distribution, all values between the minimum and maximum are equally likely to occur. For example, if there has been no information about the existing utilities on the site the value for any connections is equally likely to occur. Three conditions underlie the uniform distribution:

- ❑ the minimum value is fixed;
- ❑ the maximum value is fixed;
- ❑ all values between minimum and maximum are equally likely to occur.

Triangular distribution

The triangular distribution describes a situation where you can estimate the minimum, maximum and most likely value. Values near the minimum and maximum are less likely to occur than those near the most likely.

Figure 8.3   Uniform distribution

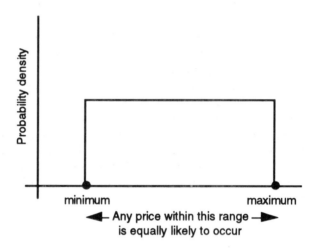

The triangular distribution is widely used because of its ease of use. However, its drawback is that it is an approximation. The approximation is a small price to pay for the benefits of using the distribution.

Figure 8.4   Triangular distribution

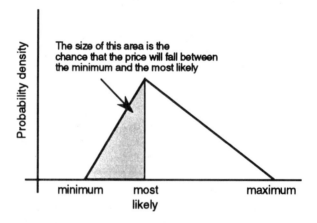

Normal distribution

The normal distribution is the most important distribution in probability theory. The normal distribution is a family of distributions, each one shaped like a bell. The bell shape spreads outwards and downwards but never quite touches the horizontal scale. The distribution uses two parameters, the mean and standard deviation. Values are distributed symmetrically about the mean and you have a reasonable idea about the variability of the data.

The normal distribution could be used when you have a good level of confidence about the most likely price. You believe that the most likely price for concrete in slabs is £67/m³. You also recognise that the price could as likely be above or below £67. The distribution uses the standard deviation which has 68% of all values within 1 standard deviation either side of the mean. Your experience tells you that there is a 68% chance that the price will be within £10 of the mean, in other words between £57 and £77/m³. In this situation, the standard deviation is £10. **Figure 8.5** shows the normal distribution with the same mean but different standard deviations. A peaked distribution will have a smaller standard deviation.

Figure 8.5   A normal distribution

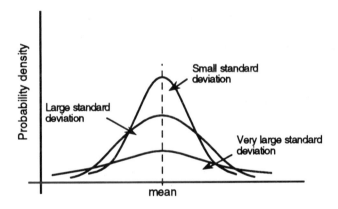

The normal distribution can also be asymmetrical (lopsided to the right or left). This situation relates to the skewness of the distribution. Standard statistical measures can be used to cope with the skewness.

Beta distribution

The value of the beta distribution lies in the wide variety of shapes it can assume when you vary the two parameters, alpha and beta. The beta distribution is discussed in more detail later.

A STEP BY STEP APPROACH TO MONTE CARLO SIMULATION

We now go on to describe the procedure in more detail for a case study. It uses historical data for a risk analysis of a cost plan. In this case, the decision maker uses the computer to generate the cost plan, having defined the distribution. A different approach would be for the decision-maker to use experience, skill and judgement to generate an estimate. Monte Carlo analysis proceeds by generating a series of simulations of a proposed project, each simulation giving a price prediction for the project. The predictions are plotted, first as a cumulative frequency curve and secondly as a histogram. There are several steps to the analysis.

Step 1

For any particular elemental category, such as the substructure, identify the probability distribution from which the price per sq m. of the gross floor area used in the prediction is taken. We shall call this the *mean unit price rate* because the rate will be derived from several projects. This is the crucial and most difficult part of the analysis, particularly since we must make some *a priori* choice of the probability distribution. It has already been shown why it is necessary to consider our sample of historical unit price rates as having been generated from some underlying probability distribution. The problem we now confront is that of choosing the appropriate probability distribution for a particular set of sample data.

Our historical sample of unit price rates gives us measures of centrality (the mean) and dispersion (the variance) which the probability distribution must also exhibit. In addition, it is desirable that the statistical distribution we use should include several other characteristics.

First, the distribution should be easily identifiable from a limited set of data, as a normal distribution is completely identified by mean and variance. This leads naturally to the second characteristic, that the distribution can be easily updated as additional historical data are introduced to the analysis.

Third, the probability distribution should be flexible, that is, capable of taking on a wide variety of shapes. We might expect that a distribution of unit price rates will be skewed to the right as in **figure 8.6(a)**.

In other words, we might expect extreme values for unit prices to be high rather than low, since the lower bound on the resource costs that underlie the calculation of unit price rate is rather more defined than the upper bound. A simple example will illustrate this point. If a vinyl tile floor finishing is considered at £15 per sq m., it is more likely to cost £30 per sq m. than it is to cost nothing!

Nevertheless, we should allow for situations in which there is no discernible skewness, or where the skewness is in the opposite direction, or where skewness is so strong as to generate a distribution such as that illustrated in **figure 8.6(b)**.

Figure 8.6(a)   Skewed probability distribution

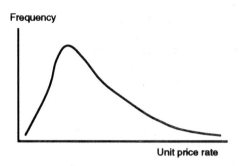

At the same time, there is no reason to believe that the appropriate probability distribution for one elemental category, for instance, substructure, need be the same as that for another elemental category, such as the external walls. It is desirable, therefore, that the probability distribution exhibits a richness in the variety of shapes it can adopt.

Figure 8.6(b)   Skewed probability distribution

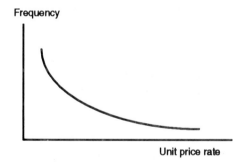

Finally, we would prefer that the probability distribution has finite end points which can be individually chosen. It is trivial to suggest that unit price rates cannot be negative. More fundamentally, we would expect that the economic decisions which unit price rates summarise - resources, design decisions, profit margins - are such as to impose lower and upper bounds on feasible or acceptable unit price rates for each elemental category. The probability distribution to be chosen should be capable of being defined within such bounds.

For the purposes of this discussion, the only set of probability distributions which adequately displays all of these characteristics is that set referred to as the *beta distributions*. Consider, for example, an alternative such as normal distribution. We have already seen in the previous chapter that where the normal distribution was used in the simulation of a life cycle cost, this distribution has the advantage of being completely identified by the mean and variance of the sample to which it is to be related. It is, therefore, quite flexible and easily updated.

The beta distribution has the equation:

$$P(x) = \frac{1}{B(p,q)} \frac{(x_{-a})^{p-1}(b-x)^{q-1}}{(b-a)^{p+q-1}}$$ (1)

$$(a \le x \le b); p,q > 0$$

where: $P(x)$ = frequency density function
a = minimum price
b = maximum price
p,q = parameters of the distribution; $p,q>0$
$B(p,q)$ = beta function

The first point to note is that the appropriate beta distribution is determined completely by the parameters p, q, a and b, parameters which themselves are generated easily from the actual data to which the distribution refers. We might take a to be the lowest value and b to be the highest value in our sample. The values for p and q are then calculated from the equations:

$$p = \left(\frac{\mu_1 - a}{b-a}\right)^2 \cdot \left(1 - \frac{\mu_1 - a}{b-a}\right)\left(\frac{\mu_2}{(b-a)^2}\right)^{-1} - \frac{(\mu_1 - a)}{(b-a)}$$ (2)

$$q = \left(\frac{\mu_1 - a}{b-a}\right) \cdot \left(1 - \frac{\mu_1 - a}{b-a}\right)^2 \cdot \left(\frac{\mu_2}{(b-a)^2}\right)^{-1} - \left(1 - \frac{\mu_1 - a}{b-a}\right)$$ (3)

Where: μ_1 = mean
μ_2 = variance

It follows that the beta distribution for a particular elemental category can be quickly and easily updated if additional historical data are added to the sample.

The shape of the beta distribution is determined by the parameters p and q, as illustrated in **figure 8.7(a) - (d)**. It can be seen that this is a particularly rich distribution. It can accommodate data which are

Figure 8.7(a)   Beta distributions for different values of p and q

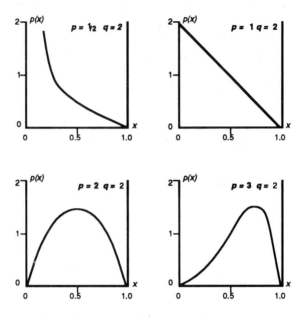

Figure 8.7(b)   Beta distributions for different values of p and q

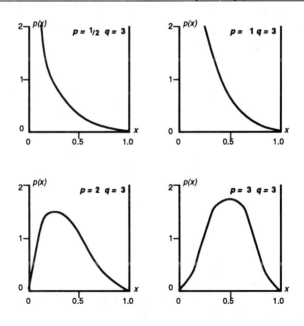

Figure 8.7(c)   Beta distributions for different values of *p* and *q*

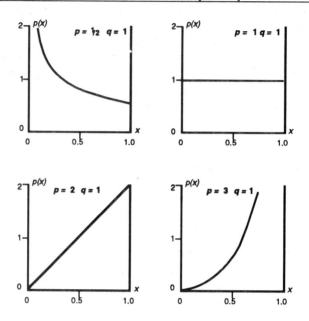

Figure 8.7(d)   Beta distributions for different values of *p* and *q*

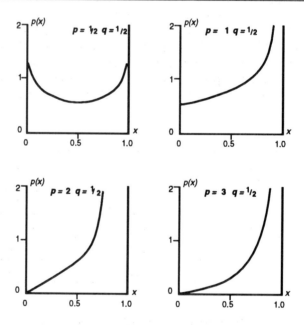

skewed in either direction, data which are not skewed at all, or data which are extremely skewed. Finally, and by definition, the beta distribution lies within finite bounds which are determined by the data to which the distribution refers.

It may be useful to illustrate the selection of a beta distribution by means of a hypothetical example. Assume that mean unit price rates for the substructure and superstructure elements have been generated from the analysis of ten completed buildings with unit price rates as given in **table 8A**.

Table 8 A   Hypothetical unit price rate

	1	2	3	4	5	6	7	8	9	10
Substructure	12.00	10.00	12.50	13.50	14.50	11.00	13.00	14.50	15.00	16.00
Superstructure	19.00	20.00	22.50	34.00	25.00	21.00	23.10	23.70	24.00	20.50

Substructure mean		Superstructure mean	
unit price rate =	£13.20	unit price rate =	£23.28
Variance =	£3.51	Variance =	£17.19

These data can be used to identify the ranges within which the mean unit price rates fall. As noted above, we approximate these ranges as running from the lowest to the highest unit price rates in our data, that is, to be £10/sq. m - £16/sq. m for substructure and £19/sq. m - £34/sq. m for superstructure. Substituting these figures for mean, variance, maximum and minimum in equations (2) and (3), we estimate the beta distributions for substructure and superstructure as shown in **table 8B**.

Table 8 B   Estimated beta distribution

Parameter	Substructure	Superstructure
a	£10	£19
b	£16	£34
p	0.83	0.45
q	0.72	1.11

The values for p and q in **table 8B** have been derived from the equations. **Figure 8.8** shows in diagrammatic form the approach for step 1 and **figure 8.9** illustrates steps 2 to 5 in diagrammatic form.

An explanation of random number sampling

A random number for computer modelling is a number generated between 0.0 and 1.0, which acts as a probability value, which in turn finds the value in the cumulative probability distribution corresponding to that probability value. Numbers chosen at random bear no relation to numbers appearing either before or after the sequence, but it must produce values in proportion to their chance of occurrence. The mechanics are governed by the shape of the probability distribution and the set of generated values will resemble the distribution that produced them.

The usual approach is to use conventional Monte Carlo sampling, but Latin Hypercube sampling is an alternative approach where the probability distribution is divided into intervals of equal probability. The approach does provide increased accuracy at the expense of more computer time and memory requirements.

Step 2

Having identified a shape for the beta distribution for each elemental category, a random number from each of these distributions needs to be generated. This is most easily achieved by using a random number generator on a computer. Each such random number is an estimate of the unit price rate for the appropriate elemental category. In our hypothetical example, the number generated for the substructure might be £12.75.

Note that this number need not equal any of the actual observations. It must, however, lie within the range (a,b), that is, £10/sq. m £16/sq. m, and have an expected value equal to the mean unit price rate, that is, £13.20/sq m.

Step 3

Multiply the random unit price rates by the measured quantities in the appropriate elemental categories for the proposed building, for example, if the measured quantity of substructure is 1000 sq m, we obtain £12.75 x 1000 = £12,750.

Step 4

Sum the results of Step 3 for each of the elemental categories used in the analysis, to give a forecast of the total project price. Store this estimate and return to Step 2. Repeat N times, where typically $N = 50, 100, 200...$, to generate N simulations of the project.

Figure 8.8   The approach for step 1

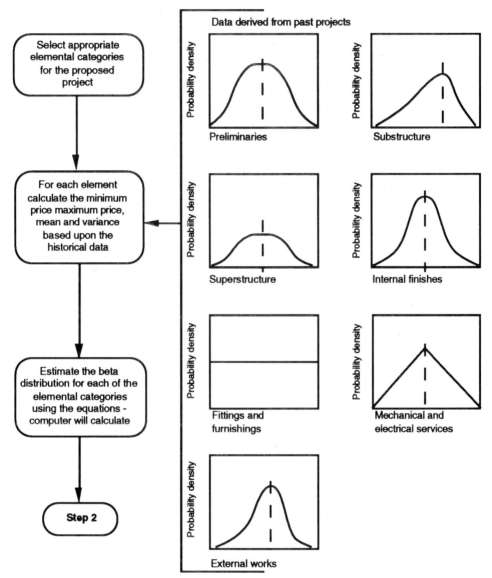

Step 5

Plot the N estimates as a cumulative frequency curve and as a histogram.
The choice of N is an immediate problem. It is preferable that N is 'large',
since this will lead to a smoother cumulative frequency curve and
histogram. On the other hand, additional simulations are not without
cost, either in computer time or manual processing.

No simple answer can be given as to the 'correct' or 'desirable' choice of N. If you have too small a number of iterations, the resulting distributions will look patchy. The larger the number of iterations, the more the distribution reflects the range of possible outcomes. A guide can, however, be obtained by consideration of the chi-square distribution, which is used to test whether a sample of data is drawn from a population with particular characteristics. The critical values of chi-square at the 5% significance level are plotted in **figure 8.10**. The degrees of freedom in this figure can be taken, approximately, to be three less than the number of intervals in the histogram produced in Step 5, while to construct a histogram we would prefer to have at least five times as many observations as intervals. Since the chi-square distribution is increasing quite quickly at 20 to 30 degrees of freedom, this would imply that we should conduct at least 100, and preferably 200 simulations in Step 4, while 500 simulations would appear to be as many as are needed.

Figure 8.9   Steps 2 to 5 illustrated

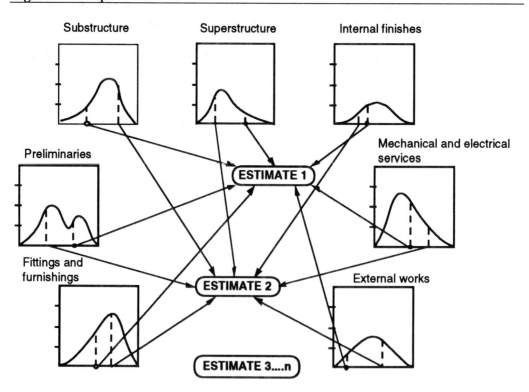

Step 6

Interpret the results carefully. Look for any inter-dependence between the elemental categories. Experience and intuition are required. The strength of correlation between two variables will show the inter-dependence. Examine the shape of the resultant distribution and the cumulative frequency diagram. The cumulative frequency distribution allows you to examine the probability of obtaining a value below a chosen value. Basically, the distribution allows you to consider the chance of the most likely price being achieved. Consider the statistics produced from the data.

Step 7

Testing the sensitivity of the data by performing sensitivity analysis on the key elements in the analysis.

Using Monte Carlo simulation on a live project

This process was undertaken for a live project and the results are presented in **figures 8.11** and **8.12**. The proposed project was a warehouse building located in an urban area. Six homogeneous completed buildings (the number of suitable buildings for which cost data were available) were selected from the cost records of the quantity surveyor concerned with the building.

Figure 8.10 A plot of the critical values of chi-square at the 5% significance level

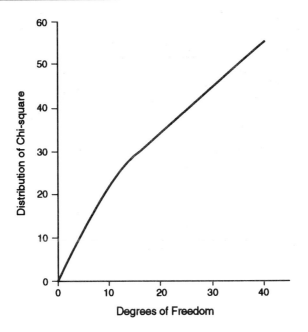

The cumulative frequency curve in **figure 8.11** indicates that 500 simulations were conducted. This curve is used to derive the probability of the price of the proposed building falling within a specified range. For example, reading along from the 250 point on the vertical axis there is a 50% probability (250/500) that the building price will be less than approximately £481,000, while reading along from the 400 point on the vertical axis, there is an 80% probability (400/500) that the building price will be less than £531,000.

The histogram in **figure 8.12** is used to supplement the cumulative frequency curve, since it indicates the 'most likely' price range for the proposed building. In this case, the most likely building price is between £470,000 and £496,000. When the simulation process has been completed and a cumulative frequency curve and histogram generated, these can be used to assess the single price forecast produced by the quantity surveyor using the unit price rates for the historical sample of buildings (see above).

Figure 8.11 Plot of cumulative frequency of forecasts generated from the cost plan

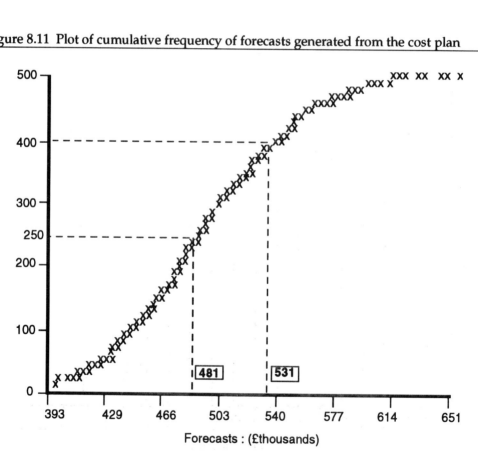

Forecasts : (£thousands)

The single price prediction for the project we have examined was £461,000. The cumulative frequency curve shows that there is a 30% probability that the tender price will not exceed the price forecast, while the histogram indicates that there is approximately a 60% probability that the tender will be within 10% of this prediction. There is no objective way of stating whether these percentages are good or bad, although it would appear that the quantity surveyor's forecast of £461,000 was slightly low. In this sense, risk analysis is by no means a substitute for the personal judgement of the quantity surveyor. However, it is a method that identifies situations in which single price forecasts should be subjected to close scrutiny.

Figure 8.12 Histogram of generated forecasts

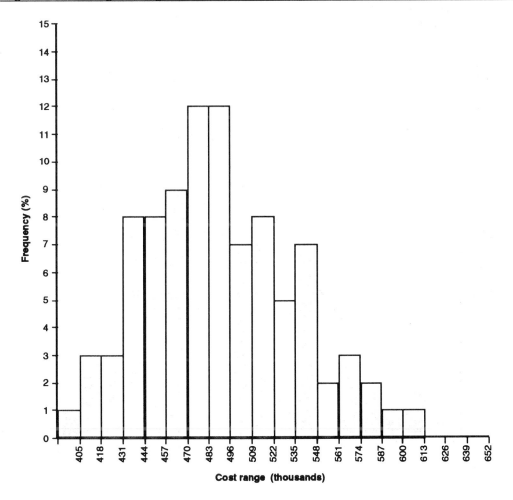

High variance in the cumulative frequency curve and histogram results from high variance in the original data. To see why this is so, it should be noted that since the total price forecasts as shown in figures 8.11 and 8.12 are generated from the summation of a series of beta distributions, that is, one distribution for each elemental category, the total price forecasts themselves are distributed as a beta distribution. It follows that the predictions will exhibit the same skew, to the right, as those of the individual elemental categories, and a variance which is related to the variances of the mean unit price rates for these categories. This implies that the more variable the historical data, the more dispersed will be the generated estimates. Hence the slope of the cumulative frequency curve will be flatter and wider the most likely price range becomes.

There is clearly no problem if the histogram is very sharply peaked and the cumulative frequency curve very steep. Difficulties arise when the total price forecasts are widely dispersed. In these circumstances the quantity surveyor has three approaches, which might be termed *macro*, *micro*, and *truth* approaches.

1. At the *macro* level, it may be that the buildings which make up the historical data set are not sufficiently homogeneous, in which case apparent consistency can be achieved by reducing the sample size to generate a more homogeneous sample. Care must be exercised in such a selection of data, to ensure that the homogeneity so achieved is not spurious, reflecting only the preconceived ideas of the surveyor.

2. At the *micro* level, it is clear that prices for certain elemental categories exhibit wider price variability than others. The substructure element, for example, has higher variability than the external doors and windows element. The quantity surveyor may, therefore, choose to examine a particular elemental category in detail to reduce the variability by more thorough analysis.

3. Uncertainty is an essential element of the forecasting process and an estimate of the uncertainty should be provided.

The result

The project was bid by five contractors. The bids received were as follows: £472,500, £496,000, £502,000, £507,000, £551,000. The lowest tender included £102,000 of prime cost and provisional sums in the bill of quantities.

QUESTIONS AND ANSWERS

Q *What happens if there is only one or two projects that can be used for the historical data set?*

A Irrespective of the simulation exercise, it would be unsatisfactory to use a single project as the sole basis of the model for a proposed project. The range of prices that could occur should always be known. Pricing is not a precise scientific exercise, it involves intuition, skilled interpretation and expert judgement. Hence, a minimum and maximum range with a mean and variance should be generated from experience.

Q *At what stage of the design process can simulation be used?*

A The technique can be used at a variety of stages. At the early design stage, it may be applied for 6 or 7 elemental categories, and so through to the detail design stage where 32 elemental categories would be used. The greater the level of detail, the more reliable the forecast is likely to be.

Q *What is better, to use a different distribution for each elemental category or use the same distribution throughout?*

A The decision rests on the professional judgement of the decision-maker. Most elemental categories will have different risk profiles. For example, the substructure will be more risky than the internal wall or floor finishes. The decision rests upon the extent of the risk for each category.

Q *If we show the minimum and maximum figures to the client, won't he look at the extreme figures and become alarmed?*

A Interpretation of the results involves professional skill. It is better to face a situation having awareness of what could happen, rather than adopting the ostrich approach.

9

CONTRACTS AND RISKS

DISAGREEMENT AND CONFLICT

All parties to a building contract will start off with the best intentions to get the work both completed satisfactorily in the agreed time and at the least expense to the owner, whilst ensuring that the general contractor and all the other specialist works contractors and suppliers make a reasonable profit. Somewhere between the beginning and the end, disagreements, disputes, disruption and delays arise which can destroy the best of intentions.

Disagreement and conflict typically arise on projects through:

❏   inadequate and defective contract documentation;
❏   inappropriate types of contract arrangements;
❏   inappropriate methods of tendering;
❏   an unreasonable burden of risk being allocated to one of the parties by the contract;
❏   unsuitable personnel for the type of project;
❏   a breakdown in personal relationships and communication;
❏   a burden of contractual risk being carried by a party who is not equipped or capable of carrying that risk;
❏   insolvency of one of the parties;
❏   interface and co-ordination problems where more than two parties are involved, for example on a project with a suspended ceiling. The general contractor, the suspended ceiling specialist contractor, the air conditioning specialist and the electrical specialist contractor must all work in accordance with a rigid construction programme;
❏   vague specially written conditions of contract or changed conditions in standard forms of contract which leads to poor interpretation;

❏ ambiguous specification clauses with decisions left at the discretion of either party;
❏ methods rather than results being specified in the specification with the contractor being responsible for providing the desired results;
❏ deficient drawings or design with discrepancies between the architectural, structural and engineering services drawings.

No matter how good the conditions of contract, disagreements will always occur on construction projects. In 99% of the cases they are resolved amicably by discussion and negotiation. Disputes are not new. When Sir Christopher Wren was designing an extension to Hampton Court Palace in 1689, it was reported that there was great pressure to get the project completed quickly. At a crucial stage of the construction a considerable portion of wall and roof fell down, causing the architect to be hauled before a court of inquiry. Like any modern day counterpart, Wren argued that there was nothing wrong with the design, but that his builders had been using sub-standard mortar while he was off-site attending to another contract - a cathedral on Ludgate Hill. He got away with it. Despite this dispute happening 300 years ago, it could have occurred on a site today : it all sounds very familiar.

THE PURPOSE OF THE CONTRACT

The purpose of the contract is to establish the rights, duties, obligations, and responsibilities of the parties and to allocate the risk. The acceptance of an obligation or duty brings with it the acceptance of a commensurate risk, which is the risk of being unable to fulfil the obligation or duty because of one's own inadequacy, incapacity, inadvertence, or error, or because of interference from outside sources or events. But with any contractual agreement the contract defines only the ground rules, the execution of the contract rests on goodwill, intent and the relationship between the parties.

A building contract is a trade off between the contractor's price for undertaking the work and his willingness to accept both controllable and uncontrollable risks. The price for doing the work partly reflects the contractor's perception of the risk involved. Lump sum fixed price contracts are the ultimate incentive for contractor performance, because of the level of risk the contractor must bear. However, the contractual arrangement should be selected on the basis of recognising who is carrying what risks.

Standard forms of building contract allocate risks between the parties by express and implied terms. However the direction of the allocation of risks can and does vary considerably. The standard forms of contract in use within the construction industry are drawn up by committees who have representatives from most of the interested parties. Employers, contractors, specialist contractors, and local authorities are all represented. The forms cover most of the risks in construction and they represent a compromise. Government forms of contract for use on government work are drawn up to suit the specific needs of government, they are concerned with public accountability. Public servants do not consider risk taking to be within the scope of their duties.

In general, public sector clients find it very difficult, for funding and political reasons, to accept uncertainty in a contract price. They prefer traditional fixed price lump sum contracts where the contractor bears most of the risks and there are few provisions for price adjustments. In the private sector however, clients, such as large development companies, are prepared to carry substantial risks on a project in return for the financial benefits and the retention of direct control of the design and construction work. Construction management contractual arrangements offer such advantages in the situation where the client contracts directly with the specialist and works contractors.

Both controllable and uncontrollable risks are important on construction projects. Controllable risks reflect, for example, variations in human performance, such as management and operative performance. Uncontrollable risks include such factors as inclement weather, the effects of inflation on costs, or ground conditions on a particular site. Contractors and specialist contractors have no difficulty in accepting the risks for controllable events, but often onerous conditions are inserted in tender documents for situations which are uncontrollable. For instance, the contractor will be responsible for ascertaining the ground conditions on the site and he will be deemed to have allowed for in his prices for excavating in any ground conditions encountered. Within a three week tender period the contractor is unlikely to be able to undertake a detailed site investigation other than some trial boreholes, hence he must price for the worst eventuality whilst at the same time balancing the commercial risk of being the lowest tenderer. To impose such risks on a contractor for uncontrollable events is short-sighted. If the ground conditions are satisfactory the client will be paying a high risk premium which could have been avoided by commissioning a detailed site investigation at the design stage.

THE FUNDAMENTAL RISKS - LIABILITY AND RESPONSIBILITY

The fundamental risks inherent in any construction project are apportioned between the client, the design team, the general contractor, the specialist contractors, and the material and component suppliers within the various contractual relationships. The risks are:

❑ adequacy of design - which party bears the risk of liability for latent defects occurring as a result of errors in design?

❑ cost of construction - which party assumes the risk of how much it will cost to build the project?

❑ liability for latent defects arising as a result of bad workmanship, faulty materials, and poor specification - which party is responsible for the inadequacy, the professional team, the prime contractor, the specialist contractors, or the material suppliers?

❑ safety and indemnification for all accidents - it is customary for one party to agree to indemnify the other for all damage and liability to third parties arising from the works;

❑ completion deadlines - which party takes responsibility for completion to the agreed deadlines?

❑ quality of workmanship and materials - which party takes responsibility for fitness for purpose and for ensuring the quality is acceptable?

Underlying all the above items is some form of liability and responsibility. The law regarding liability is complicated and it is beyond the scope of this book to consider liability in detail, particularly how it affects design professionals. All professionals owe a duty of care and if a latent defect occurs as a result of negligence by the architect or engineer in their design, then the client may seek recovery for the damage suffered. The complication arises when a defect occurs as a result of the failure of a specialist contractor's design detail. Frequently, a conceptual design detail is given with a performance specification. In this case the specialist contractor will submit the detail design, as a shop drawing, to the architect for his approval that the design conforms with the general design concept. If approval has been given and a latent defect occurs as a result of the inadequate design detail prepared by the specialist contractor on the shop drawings, then the client must ascertain why and how the defect occurred and whether the architect was professionally negligent in approving the shop drawings. Theory and reality part company at this point. In practice, the reason for defects occurring is rarely straightforward, and the client has to 'join' a number of parties in an action to ascertain who was responsible and who should pay for the damage suffered.

TRANSFERRING AND ALLOCATING THE RISK IN THE CONTRACTS

The cost implications of certain construction risks is very high. Risk sharing becomes virtually impossible in practice when one party to the contract attempts to shift the majority of the risk to the other party through one sided contract language.

It is not just the owner who passes the risk to the contractor, frequently the risk is transferred to the specialist contractor by the use of indemnity clauses or 'hold harmless' clauses. Considerable care needs to be taken in the drafting of hold harmless clauses. In the past the courts have been unsympathetic to parties trying to contract out of liabilities. A good example of this is in the Sale of Goods (Implied Terms) Act 1973, and the Unfair Contract Terms Act 1977.

Furthermore there is no point in passing a risk, through contract conditions, on to a party who is ill equipped to bear that risk. For instance, the contractor will be subject to liquidated and ascertained damages for delays to completion of the works. Frequently the amount of the liquidated and ascertained damages in the main contract is inserted in the specialist contractor's and sub-contractor's domestic contract agreements. To take an extreme case, it is unrealistic to impose liquidated damages of £10,000 per week on a three man shuttering specialist contractor. The company is unlikely to have the financial resources to meet such costs even if it could be proved that they were responsible for delaying the project.

Most of the standard forms of building contract are well understood and it is recognised who is carrying the risk. When a special form contract is drafted for a project there is a serious risk that much is lost, distorted or overlooked in the process of translating the wishes of the parties into legal language. However there are instances when clients will want to amend the standard form of contract to meet their particular requirements. For instance, on quality, the contract states that 'all materials, goods and workmanship shall, so far as procurable, be of the respective kinds and standards described in the contract bills'. The specification or bill of quantities will describe the quality required. The standard form of contract requires that if defective workmanship is found it shall be removed or otherwise remedied. No 'assurance' is offered by way of quality control and other checks: obligations in this respect have to be written into the contract as a special requirement.

The allocation of risks is inherent to the apportionment of legal liabilities. It is at this stage of apportionment that the principles of control are significant. The adverse events with which the risks are associated may be caused by the employer's or the contractor's acts or defaults, or be caused by natural hazards and other unforeseen conditions. The employer or the contractor who is allocated the risk will be required to accept the consequence flowing from the occurrence of the adverse events. A party contracting in the light of expectations must make up his mind whether he is prepared to take the risk of these expectations being disappointed. If not, then he will refuse to contract. It is also possible that the consequences may involve a conferment of certain rights on the aggrieved party. So a contract such as the Joint Contracts Tribunal Standard Form of Building Contract is an agreement to translate responsibilities of the parties for specific risks into a set of liabilities and rights conferred on both the employer and the contractor.

However, the risks borne by the employer and the contractor are not limited to those allocated by the contract. Certain risks are allocated by operation of law. The more notable examples are the rules governing the employer's liabilities to his workmen and the rules governing the frustration of contracts. There is a risk that a workman may be injured in the course of his employment as a result of the contractor's default in failing to provide safe equipment or a safe system of working. There is also the risk that some intervening event (beyond the control of the employer and the contractor) may alter the nature of the whole contract such that one of the parties is able to successfully claim that the contract is frustrated. In such circumstances, there are adverse consequences to be borne by the party who is allocated the risk. The allocation arises not by contract but by the operation of established rules of law.

THE PRINCIPLES OF CONTROL - THE THEORY

A very significant factor of risk allocation is the element of control available to the parties who are required to bear the risks. A rational framework of risk allocation will require the party, who is in a better position to control the occurrence of adverse events, to bear the risks associated with such events. The employer, who is able to control the date of possession of the site, is the party who should bear the risk that possession of site on the date specified could not be affected. Similarly, the contractor who is able to control the manner in which the work is to be carried out, is the party who should bear the risk of injury that may be caused to the public by the contractor's workmen.

This element of control can be extended to include situations when the control is not as simple and direct as that cited. Control can be imputed to the employer where causation of damage is due to the actions of the employer's authority. Under the traditional contractual arrangements, the responsibility for risk is not necessarily associated with control. For example, a design requirement of the architect may not be in accordance with a particular statutory regulation and yet it is the contractor who has to bear some of the loss arising out of the breach. The contractor is considered to be a professional and if he failed to exercise reasonable skill and care by not checking the drawing prior to construction work, then he will carry the liability.

Naturally, the element of control is associated with a state of circumstances and cannot be looked at in isolation. So, when that state is altered as a result of some intervening act, it cannot be said that the original risk bearer is still in control. He should then be allowed to transfer any adverse consequences of the intervening act to the party causing the act. For instance, the general contractor is in control of the execution of the construction works on site and is therefore the party expected to bear the risk that the works may not be completed on time.

However, if the employer interrupts this control by suspending execution of part or the whole of the works, then the risk for any adverse effect must be transferred to the employer.

The difficulty with a building project is that it is unique and involves a vast number of parties. This is a recipe for disaster, but generally, it works satisfactorily. With a conventional lump sum bid the contractor and his specialist contractors are given a month to price a project that has been at the design stage for several years. The underlying assumption is that the project has been fully designed and specified prior to obtaining tenders; however in reality this is rarely the case. In the UK the detailed design is developed as the project is under construction. The contractual arrangement has to be flexible to cope with this requirement and various different contractual arrangements have developed over the past twenty years.

THE CONTRACTUAL LINKS

The variety of contracts that exist between the various parties in the design and construction process is bewildering. Within each contract the liabilities and responsibilities are determined. Most importantly each contract defines the risks being carried by the parties.

Each of the parties will decide what risks they are prepared to retain and which to transfer. When the employer-general contractor contract is a standard form of building contract, then there are standard forms of contract that should be used for the general contractor-specialist contractor contractual arrangements. Frequently the major contractors use their own forms of contract, to contract with specialist contractors, which contain onerous conditions and impose greater risks on the specialist than is reasonable. For instance, the amount of liquidated and ascertained damages, payable as a result of failure to complete the project on time caused by the failure of the specialist contractor, includes the main contract liquidated damages and damages suffered by the general contractor.

RISK AVOIDANCE BY WARRANTIES AND COLLATERAL WARRANTIES

The contractual links have been further extended over the past decade by the use of warranty agreements; more recently collateral warranties have been sought.

Under English law the advantages of strict contractual liability are offset by the requirements of privity which confine the benefits and liabilities to the immediate contracting parties. As a result, an employer may not claim under the law of contract against a sub-contractor with whom he is not in contract. However, this does not affect the position in tort where there might be tortious liability.

Under the Standard Form of Building Contract there is provision for a nominated sub-contractor to sign an employer-sub-contractor agreement where there is provision that the sub-contractor has exercised and will exercise all reasonable skill and care in the design of the sub-contract works in so far as they have been designed by him.

The law relating to warranties in the construction industry is not clear. It has been further complicated by the introduction of collateral warranties.

In commercial and industrial developments the client is a developer who will sell to a pension fund or financial institution who then let the premises to tenants on a full repairing and insuring lease. If a latent defect occurs in the property five years after completion the tenant could be in the position under the lease of having to pay for the repair of the defect. Simply, the collateral warranty is assignable to another party and it binds the parties in a contractual obligation as shown in **figure 9.1**.

This is another example of risk transfer. The pension fund and the tenants have the security of being able to pursue the company responsible for any negligence or shortcoming in design, specification or workmanship.

Figure 9.1   Contract and warranty arrangements

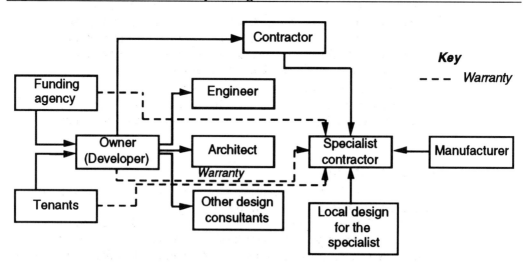

RISK TRANSFER BY SURETY BONDS

In law a surety is a party that assumes liability for the debt, default, or failure in duty of another. A surety bond is not an insurance policy, it is the contract that describes the conditions and obligations of such an agreement. Insurance protects a party from risk of loss, while suretyship guarantees the performance of a defined contractual duty.

Contract bonds are three party agreements that guarantee the work will be completed in accordance with the contract documents and that all construction costs will be paid. Regardless of the reason, if the contractor or sub-contractor who is bonded fails to fulfill their contractual obligations, the surety must complete the contract and pay all costs up to the face amount of the bond.

Contract bonds are required in the USA by law on public sector work on projects of over $25,000 in value; whereas a substantial proportion of private sector work is not bonded. In the UK, contract bonds for building work are not common, but they are being increasingly used to bond the sub-contractor to the general contractor for completion. The bond does not serve to replace the honesty, integrity and ability of the sub-contractor. Bond or no bond, an inferior sub-contractor means trouble. The bond will afford the general contractor some measure of protection against financial loss directly attributable to a particular sub-contractor.

Requiring a bond is a way of risk transfer, where the risk of financial failure is transferred to the surety. The surety company is normally an insurance company, a bank, or a company specialising in providing sureties.

The types of surety bond in use are:

❑ bid bond ensures the contractor will stand by his tender bid;
❑ performance bond ensures that if the contractor defaults, the project will be completed in accordance with the terms of the contract. All performance bonds have a face value which acts as an upper limit of expense the surety will incur;
❑ labour and material payment bond protects the employer for labour and material used or supplied on the project. It protects against liens being filed on the project by unpaid parties to the work.

The types of contract

Table 9A shows in simplified form who carries the risk in the various types of contract used in the building industry.

The professional team will select the contractual arrangement that offers most advantages to the employer. On larger value commercial projects there has been a trend towards the management fee arrangements where the general contractor plays an active role as a member of the design team. In the early stages of management contracting the general contractor carried very little financial risk as his fee was based upon a percentage of the overall construction price. Times have changed and often management fees are fluctuating, based upon a target price, with some employers demanding a guaranteed maximum price from the management contractor.

Under the management fee contract the contractor guarantees that the project will be constructed in accordance with the drawings and specification and the cost will not exceed a total maximum price; sometimes referred to as the upset price. If the cost exceeds the assured maximum, the contractor pays the excess. An incentive for the contractor to keep costs below the maximum is sometimes provided by a bonus clause stating that the contractor and the employer will share any savings.

The move towards requesting a guaranteed maximum price on management fee contracts is putting the contractor back into the position of lump sum bidding without the same degree of control. Risk and reward go hand in hand but the contractor has a high financial risk exposure and a relatively low return with the guaranteed maximum price; the management fee must reflect the risk exposure. However, the fee arrangement contracts do have advantages where work can start on site before the design is complete.

Table 9 A   Who carries the risk in the various types of contract

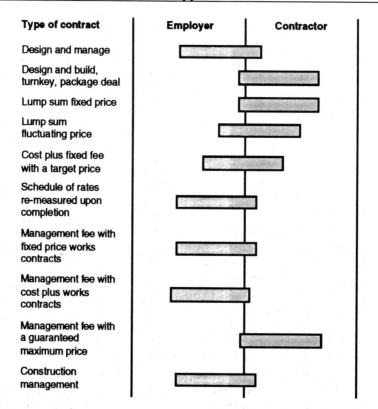

CONTRACTS AND RISK TACTICS

Figure 9.2 shows a list of sources of risk from the viewpoint of the general contractor at the tender stage. The aim of the list is to identify the risk and the offsetting tactic needed to ensure the risk is adequately covered.

Figure 9.2   Source of risk and offsetting tactics

RISK SOURCE	Contract conditions	Tender documents	First payment	Advance payments	Currencies of payments	Payments to sub-contractors	Limit to penalties	Fixed price quotes	Insurance (various)	Indemnification	Income guarantees	Taxation conditions	Contingency/Risk allowances	Investments
1. Client cancel	●			●					●					
2. Client fail to pay	●			●					●					
3. Client suspends	●													
4. 'Force Majeure' incident	●								●					
5. Contractor cancels	●			●										
6. Contractor suspends	●													
7. Client acceptance delay		●	●											
8. Client start delay	●			●										
9. Suppliers start delay	●						●		●	●				
10. Transport start delay	●								●					
11. Contractor start delay							●		●					
12. External start delay	●	●	●						●					
13. Client duration extension	●				●				●					
14. Suppliers duration extension	●						●		●	●				
15. Transport duration extension	●								●					
16. Contractor duration extension							●		●					
17. External duration extension	●								●					
18. Base cost estimation error								●					●	
19. Escalation estimation error									●					●
20. Exchange rate excalation error	●				●	●								
21. Investment income estimation error											●	●		
22. Investment error												●		
23. Supplier performance failure							●		●	●				
24. Contractor performance failure									●					

Source: Thompson and Perry.

The contractor should be aware at all times of the extent of his risk exposure - the one thing that all contractors hate is a surprise.

The figure shows that the contractor has examined the conditions of contract and the tender documents to see the sources of risk likely to affect his performance. For example, he has identified that the design has not been fully developed and he runs the risk of claims for loss and expense from the specialist and works contractors, caused by the lack of information. As a general contractor on a lump sum contract he would also be able to claim for any disruption caused to the progress of the work. However, at the tender stage having identified the risk he can absorb it by adding a risk premium to the prices and he can include special conditions in the contracts with the specialist contractors such that they must ensure the information is adequate for construction purposes; thus transferring some of the risk to the specialist contractors.

Basic factors relating to risk in contracts:
- ❏ **what is the exposure inherent in the contract**
- ❏ **who is most capable of handling that exposure**
- ❏ **who has the responsibility for that exposure**
- ❏ **who has the power to make sure that responsibility is carried out**
- ❏ **what has been done to take account of the uncontrollable risks**
- ❏ **to what extent have the risks been transferred**

10

A CASE STUDY OF AN
OIL PLATFORM

This case study has been prepared by Euro Log Limited and relates to the
Phillips Ekofisk Elevation Project which tackled the subsidence of major
offshore structures in the Norwegian sector of the North Sea.
(We are grateful to Phillips Petroleum Co. Ltd., and to Euro Log
Limited of Teddington, UK - Project Risk Management Consultants - for
permission to reproduce this case study.)

A PRACTICAL APPLICATION OF RESOURCED SCHEDULE RISK ANALYSIS

This chapter shows how risk management techniques were used on a major
North Sea project. The aim is to give an insight of how such techniques can
be used in practice.

The Ekofisk Elevation Project had many unique features, not the least
of which was the need to execute the entire project, from design through to
the actual platform elevation, within a span of less than 16 months. As a
result, planning and project management became key elements not only to
maintain control over the project but also to ensure that it was performed
within this tight timeframe.

Conventional deterministic planning was employed to a great extent,
but could not provide key elements required for this project; a reliable
prediction of the probability of achieving the goal of jacking the Ekofisk
platforms within the 1987 summer season and the preservation of an
acceptable certainty level on that goal. To satisfy these needs,
probabilistic scheduling was employed.

The purpose of this case study is to show how schedule risk analysis
actively contributed to the management of a live project.

The elevation of Phillips Petroleum's Ekofisk platforms was a
complex project, of unusual urgency and prominence, and it was unique.

Phillips adopted a risk analysis and management approach to direct its execution.

Following a background to the project, this case study describes the probabilistic plan (model), including the input, the level of detail, data collection, and how Phillips used the results to drive the Elevation project.

BACKGROUND

The Ekofisk Field is strategically situated in the Norwegian sector of the North Sea. Its facilities have become a key element in Norway's oil and gas transportation, handling 300,000 barrels of oil per day and Norway's entire gas exports to Europe. The plan of the field layout is shown in **figure 10.1**.

Figure 10.1 Ekofisk Field layout

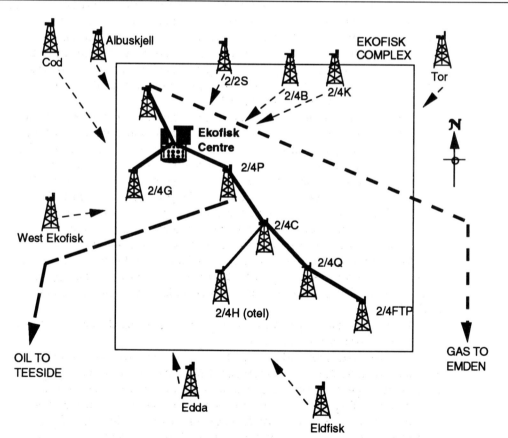

In the early 1980s, subsidence was detected at the central complex of platforms and, by the winter of 1985/86, it was clear that the full extent of the subsidence would be about 6.0 metres. At that level of subsidence, platforms would have to be abandoned unless fundamental modifications were made. Within two years, safety would require production shutdown in severe weather.

The platforms critically affected were six steeljacket structures interconnected by bridges and extensive production piping. A seventh platform, the concrete 'Tank', is at the hub of this piping and was not to be modified in this project.

The decision was made to extend the six jacket structures by 6.0m. To avoid the enormous costs and long term implications of repeated production shutdown, production piping between the six platforms was to remain intact. The platforms were to be raised simultaneously reducing pipework modifications to those at the Tank, and to riser connections.

The entire shutdown period was to be 28 days, coinciding with a 2-week maintenance shutdown in summer to minimise the loss of sales.

The overall period from start of design to the restart of production was extremely short at 16 months. The design contract was awarded in April 1986. The shutdown constraints prescribed a platform elevation period in the summer of 1987. In the intervening period, the computer-controlled jacking system for 120 jacks, jacket strengthening, 2500 tonnes of temporary support steel and a monorail installation system had to be designed, manufactured, installed and commissioned.

The scope of work and logistics for such a time frame were enormous. The project was unique and almost all of the pre-shutdown installation work was in the splash zone, and so subject directly to North Sea weather conditions. These constituted major uncertainties which Norwegian Authorities, Phillips and their partners and customers had to reconcile with their own commitments. In addition, the production revenue of several oilfields was at stake, as were the tariffs for transportation.

The Phillips project team had to reconcile uncertainties and complexity with two of the basic needs of project management: maintaining a realistic forecast of completion; and effectively directing their limited resources to expedite the project. Above all, their objective was to ensure that platforms were elevated before the onset of winter 1987/88.

The costs of the project and consideration of physical safety made the schedule of paramount importance.

The risks surrounding this project were enormous and so schedule risk analysis was introduced in September 1986, midway through design when serious slippage was becoming apparent against the first milestones.

The prevailing opinion in the industry favoured completion during or after winter 1987/88 because of the slippage, emergent design and contractual problems, and the fact that the 'real' problems - weather, logistics, finding and putting to work a large labour force were yet to be faced.

The first objective of the risk analysis was, therefore, to determine a completion date in which Phillips and their partners and customers could have confidence and to answer the following questions of the target date:

❑ was it achievable?
❑ what was the probability of achieving it?
❑ what was the possible extent of overrun?

The second objective was to provide a means by which to co-ordinate and drive the project, answering the following questions:

❑ where, considering the uncertainties, were the critical areas?
❑ what were the reasons for criticality?
❑ what could be done in-house to improve the likelihood of early completion?

THE MODEL

The first task was to establish the probabilistic plan, or model, of the project.

Comparison with deterministic plan

The model was to resemble a conventional plan of being network-based. The model was so called in that it had fully to represent what could happen to the constituent activities. Durations, resources and, where appropriate, logic were to be variable according to the uncertainties recognised by the modeller. Instead of, say, a duration being represented by a single value it was to be represented by a range from minimum to maximum.

In addition, constraints and project interference (such as weather) were built-in such as to represent the way an activity would actually respond to the constraint or interference.

Data

As in conventional planning, the first requirement was to establish the project structure and develop a network. The entire scope of outstanding work, including the pre-shutdown and shutdown periods, was represented by 1200 activities.

To establish activity durations, estimates were calculated from various sources, considering resource constraints where applicable. Because of the unconventional nature of many of the tasks, these had to be estimated from first principles (for example, how many men for how long), using information from project personnel. Little contractor information was available.

The repetition of activities permitted a systematic approach so that, where necessary, estimates were made at a very detailed level. Many activities represented subnets of up to 14 tasks. Construction procedures were examined and an optimised approach was adopted reflecting information gained at a workface level of detail.

Constraints imposed include:

Resource limitations
- ❏ men per workface
- ❏ men per platform
- ❏ trades availability
- ❏ shifts per day

Safety
- ❏ hotwork restrictions
- ❏ work overhead
- ❏ interruptions to permit execution of priority activities.

Weather
- ❏ wind
- ❏ waves

Weather

Sea conditions were the main consideration. Work zones were defined according to height above sea level. Correspondingly, work in these zones was interrupted in sea conditions exceeding the respective thresholds and would only recommence given acceptable durations of good weather. This combination of wave height and persistence defined the good and bad weather conditions.

Seven weather models were developed to reflect conditions actually experienced in the field. These included models for three wave heights with alternative forecast periods of workable weather for each sea condition. In addition, there was one model of wind speed.

The source data was six years of Ekofisk field records. Each good and bad weather period over these six years was individually reflected in the models.

The weather data showed the unpredictability of the weather. Even in the month of February, the most vulnerable activities at one metre above sea level were shown to be executed without weather delays on occasions. On other occasions, delays of over a month were indicated.

It was important in securing the confidence of Phillips management and customers that this major concern was seen to be properly modelled. A global approach, using average down times per month to extend activity durations would not have reflected the schedule uncertainty that was known to exist.

Project variables

The significance of modelling, versus conventional planning, is that it permits the full use of all the information available.

No information is discarded by reducing a range of data to a single value. For any aspect of this model, say an activity duration, the range and nature of data available is reflected as a distribution of values. Variability was due to uncertainty in three main areas: estimating, potential problems and weather interference.

Uncertainty in estimating was generally the most significant. Factors of 2:1 in comparing estimates of the same work from two estimators was not uncommon. Variability was established by using various sources (different estimators, different norms, site experience), considering historic variations in contractor performance, and considering uncertainty of scope.

The second area of uncertainty - potential problems - also reflected information gained by examining the work at hand, both independently and with project personnel. The impact of each problem was then estimated. For example, the occurrence of jack failure during offshore commissioning was anticipated. The information modelled was in answer to questions such as:

❑ What are the chances of failure?
❑ Where would it occur?
❑ What would be involved in changing-out a 24-ton, 9m-long hydraulic jack, replacing it, hooking-up and recommissioning?
❑ How many men, how many days?

The shapes of distributions were determined by the nature of the information available and varied from uniform, to triangular, to free-format cumulative continuous distributions.

Processing of data

The Ekofisk project was a major scheme and the software had to be capable of processing the model with large amounts of data in various formats.

On the Elevation project, PROMAP V software was used on a Prime mini-computer.

The pre-shutdown model comprised approximately 950 activities and 22 resources with constraints of resource limitations and activity priority.

In addition, there were 400 weather-sensitive activities affected by seven interacting weather models.

Confidence in the data

Without Phillips confidence in the input to the model, the results would have been devalued. Both the input and the results had to withstand scrutiny of individual project specialists. The project team was, therefore, heavily involved in the research and review of data.

To ensure this involvement, a cycle of collective briefing, model presentation and result reviews was followed in conjunction with individual interviews.

INITIAL RESULTS

There were two project targets.

☐ Preparations for elevation were to be completed ready for production to be shut-down on 15th July 1987.
☐ The production shutdown was to be complete within 28 days.

Typically, the schedule forecast results were presented in the form of Time Summary graphs as shown in **figure 10.2**. The schedule results shown in the Time Summary graph in figure 10.2 are summarised in **table 10A**.

These results indicated that the Ekofisk Elevation could indeed be executed in the summer of 1987, against the prevailing climate of opinion. It is notable, firstly, that the spread of dates was not as wide as the perceived uncertainty had suggested. Secondly, even the pessimistic results did not extend into the winter period.

Table 10 A Summary of schedule results

Shutdown	Probability of achievement			
	1%	60%	80%	99%
Date 1987	9 Jul	5 Aug	10 Aug	11 Sep
Duration (days)	25	30	32	48
Target	Probability of achievement			
15 July	2%			
28 days	15%			

Figure 10.2 Time summary graph - shut down

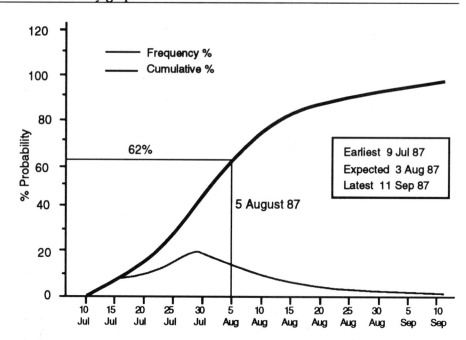

In spite of the apprehensions felt about this project at the time of modelling, the results were considered sound as a result of the approach taken. They relieved the project team of much client pressure and endorsed Phillips' view that the project should continue to be driven by the 15th July 1987 target for shutdown.

The results were contingent on the assumptions and requirements set out in the model, to which Phillips responded positively. Their success in recognising and taking action in the risk areas was borne out by the achieved outcome of the project.

The actual date of shutdown was 10th August 1987 in spite of some major re-design. The duration of shutdown was 25 days, following a reduction in scope, investment in equipment and extensive planning.

The results also illustrated a means by which the project team could best direct their efforts. Because of the variables incorporated in the model not one but all the potentially critical paths were identified as illustrated in the simplified Criticality Diagram (**figure 10.3**). These relatively few areas, out of a 950 activity pre-shutdown network, would become the focus of management attention.

By referring to the support information for data input, the causes of criticality were identified and mitigating action was taken at early stages.

Figure 10.3 Criticality diagram

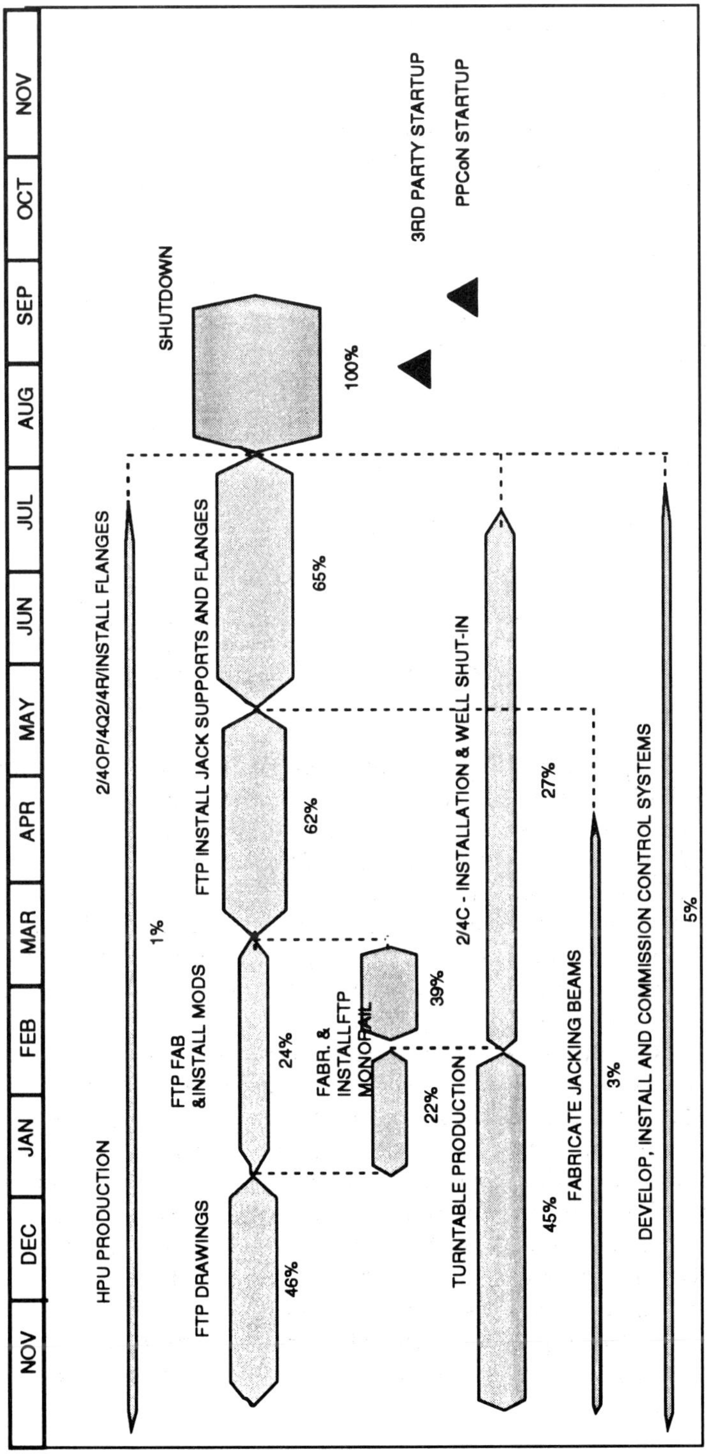

CONCLUSION

This case study illustrates the application of resourced schedule risk analysis in actively supporting the direction of a live project.

By the depth of insight it provided to the project, the risk analysis gained the confidence of partners, customers and project team alike. Its effectiveness throughout the project rested on the sustained reliability of the model and the speed of response to new information.

One of the strengths of the risk analysis was that it related simultaneously to the workface levels of detail and to the high-level overview required by most senior management.

Operated by a two-man team, it was cost effective and complemented Phillips' approach to managing the project with a compact team of key personnel.

Phillips Petroleum management considered that, without the risk analysis and risk management approach, 'it would have been difficult, if not impossible, to meet the project deadline'.

REFERENCES AND BIBLIOGRAPHY

REFERENCES

Baker A J (1985) — *Information about Risks and Uncertainties under a Weak Ordering of Preferences*, University of Leicester, Discussion Paper No 43.

Baker A J (1986) — *Valuing Information about Risks*, University of Leicester, Discussion Paper No 51.

Bennett J and Ormerod R N (1984) — *Simulation Applied to Construction Projects*, Construction Management and Economics, 2, 225-263.

Chapman C B (1979) — *Large Engineering Project Analysis*, IEEE Trans Eng Management, EM-26, 78-86.

Cooper D F, MacDonald D H and Chapman C B (1985) — *Risk Analysis of a Construction Cost Estimate*, Project Management, 3, 141-149.

Flanagan R, Meadows J, Norman G and Robinson G (1987) — *Life Cycle Costing: an application to building finishes*, Surveyors Publications Ltd, London.

Flanagan R and Norman G (1983) — *Life Cycle Costing for Construction*, Surveyors Publications Ltd, London.

Hertz D B and Thomas H (1983) — *Risk Analysis and its applications* Wiley & Sons, London.

Hayes R W, Perry J G and Thompson P A (1983)	*Management Contracting* Construction Industry Research and Information Association.
Perry J G and Hayes R W (1985)	*Risk and its Management in Construction Projects*, Proc Instr C W Engrs, 78, 499-521.
Peters, T. (1988)	*Thriving on Chaos*, Macmillan (London) Ltd.
Thompson P A and Perry J G, (1983)	*Engineering Construction Risks - A Guide to Project Risk Analysis and Risk Management*, Thomas Telford.

BIBLIOGRAPHY

Abramowitz A	*Risk Management*, American Institute of Architects
Armstrong A (1986)	*A Handbook of Management Techniques*, Guild Publishing, London.
Barnes M (1983)	*How to Allocate Risks in Construction Contracts*, Project Management, Vol.1 No.1, February 1983.
Borch K (1969)	*A Note on Uncertainty and Indifference Curves*, Review of Economic Studies, 36.
Brooks, H F (1988)	*Risk management & Insurance for Equipment Lessors*. American Association of Equipment Lessors.
Charette R (1989)	*Risk Management*, McGraw-Hill Publishing Co.
Cooke S and Slack N (1984)	*Making Management Decisions*, Prentice-Hall International, London.
Cox, M B (1991)	*Risk Management for the Department Head - An Integrated Approach*, Cox Publications.

Darvish T and Eckstein S (1988)	*A Model for Simultaneous Sensitivity Analysis of Projects*, Applied Economics, 20 No.1 .
Devine P et al (1985)	*An Introduction to Industrial Economics*, 4th Edition, Allen and Unwin.
Feldstein M (1969)	*Mean-Variance Analysis in the Theory of Liquidity Preference and Portfolio Selection*, Review of Economic Studies, 36.
Feldstein M (1978)	*A Note on Feldstein's Criticism of Mean-Variance Analysis: A Reply*, Review of Economic Studies, 45.
Fisher C (Editor) (1991)	*Risk avoidance in construction contracts*, HLK Global Communications Inc., USA,
Flanagan R and Norman G (1983)	*Life Cycle Costing for Construction*, RICS.
Flanagan R, Shen L and Mole K (1988)	*Life-cycle Costing and Sensitivity Analysis*, Proceedings of the British-Isreali Seminar on Building Economics, Haifa, May 1988.
Grubel H (1981)	*International Economics*, Irwin
Harpster, Linda, Veach M S and J D (1990)	*Risk Management Handbook for Health Care Facilities*, American Hospital Publishing Inc.
Hayes, H (1986)	*Risk Management in Engineering Construction*, State Mutual Book & Periodical Service.
Hayes R, Perry J, Thompson P and Willmer G (1987)	*Risk Management in Engineering Construction*, SERC.
Head, G L (1984)	*Risk Management Process*. Risk Management Society Publishing Inc.
Koutsoyiannis A (1982)	*Non-Price Decisions , The Firm in a Modern Context* , Macmillan.
Lenz M Jr and Ralston. (1988)	*Risk Management Manual* , Merritt Co.

Levick D E (1988)

*Risk Management & Insurance Audit Techniques.,*Shelby Publishing Corporation.

Machina M and Rothchild M (1987)

Risk in Eatwell et al "The New Pelgrave : A Dictionary of Economics", 4, 201-206.

Mayshar J (1978)

A Note on Feldstein's Criticism of Mean-Variance analysis, Review of Economic Studies, 45.

Meyer J (1987)

Two-Moment Decision-Models and Expected Utility Maximization, American Economic Review, 77, No. 3, June 1987.

Minor J K and Vern B (1991)

Risk Management in Schools, Corwin Press Inc.

Morrison D A (1991)

Risk Management and Loss Control Manual for Local Government, Sound Resource Management Press.

Naylor T and Vernon J (1969)

Microeconomics and Decision Models of the Firm, Harcourt, Brace and World.

Papageorge T E

Risk Management for Building Professionals. Mans, RS Company Inc.

Philip Allan. Porter C (1981)

Risk Allocation in Construction Contracts, M.Sc Thesis, University of Manchester.

Singh G H and Kiangi G (1987)

Risk and Reliability Appraisal on Microcomputers, Chartwell-Bratt (Publishing) Ltd.

Temple D M (1981)

Risk Management & Insurance for Commercial Bankers, Robert Morris Associates.

Thornhill W T (1990)

Risk Management for Financial Institutions: Applying Cost-Effective Controls & Procedures, Bank Administration Institute.

Tobin J (1969)

Comment on Borch and Feldstein, Review of Economic Studies, 36.

Ward C W R (1979)

Risk Analysis and its Application to Property Development, PhD Thesis, Reading.

Warren D (1985)

Risk Management Guide, Risk Management Society Publishing Inc.

Wasserman, Natalie and Phelusm (1985)

Risk Management Today: A How-to Guide for Local Goverment, International Management Association.

Williams, CA and Heins R M (1989)

Risk Management and Insurance, Witherby & Co Ltd., UK.

INDEX